ENERGY POTENTIAL

TOWARD A NEW ELECTROMAGNETIC FIELD THEORY

ENERGY POTENTIAL

TOWARD A NEW ELECTROMAGNETIC FIELD THEORY

by Carol White

with

GRAVITY, ELECTRICITY, AND MAGNETISM

and

A CONTRIBUTION TO ELECTRODYNAMICS

by Bernhard Riemann

Translated from the German by
James J. Cleary, Jr.

Campaigner Publications, Inc.

university editions

COGITO ERGO SUM

New York

The cover depicts vortex filaments from the current sheath in the plasma discharge from a theta pinch experimental fusion machine. These filaments evidence the natural tendency of a plasma toward organized and self-differentiated structure.

Photo courtesy of Dr. Winston Bostick.

Energy Potential: Toward A New Electromagnetic Field Theory

 Printed in the United States of America. For information address Campaigner Publications, Inc., University Editions, 231 West 29 Street, New York, New York 10001. Editor: N. Spannaus; Associate Editor: K. Stevens; Book Designers: D. Squires and E. DeGroot; Graphics: A. Yue, G. Genazzio, and G. Baier.

International Standard Book Number: 0-918388-04-X
Library of Congress Catalog Card Number: QC665. E4W45

CONTENTS

Note to the Reader

While a book such as this is most easily understood by a reader who is familiar with electromagnetism and related fields, it is meant as well as an introduction for the nonscientific layman. If he will read through the book without worrying unduly about mathematical formalisms which are unfamiliar to him, he will have the necessary overview with which to learn them by subsequent study. *I would like to acknowledge with appreciation the collaborative efforts of Uwe Parpart, Dr. John Schoonover, and Kenneth and Molly Kronberg without whom this book would not have been possible.*

ENERGY POTENTIAL

TOWARD A NEW ELECTROMAGNETIC FIELD THEORY

by Carol White

Dedication

Michael Faraday is rightly acclaimed for establishing the empirical science of electromagnetism. It was he who first understood the possibility of the electric motor; it was he who first induced electrical current; and it was he who discovered the quantitative laws which determine the interplay between electricity and magnetism. He was not "merely" a practitioner. His experimental work was shaped by his determination to discover the overriding laws of a unified field theory which would describe the propagation of electromagnetic, light, and gravitational effects coherently.

Nonetheless, he was tragically limited in scope. He was virtually a self-educated man. It is hardly to be wondered that he could not understand the mathematical accomplishments of the nineteenth century which were the necessary basis for such a unified field theory. As a result, his work was left to other men to shape mathematically, most notably to James Clerk Maxwell. Faraday's shortcomings, the results of his ignorance, were reified by these epigones. His field conception was trivialized to the familiar notion of lines-of-force. His search for a unified field theory was reduced by Maxwell to the search for an agreeable model on which to hang experimental results.

Faraday was a practicing member of a small Protestant sect, the Sandemanians, who celebrated the glory of God by communal Sunday dinners, preceded by a solemn ritual during which the members would wash each other's feet in token of their humility before God. It was appreciatively remarked by all who met him what a humble man Faraday remained despite his fame. That same "humility" characterizes his experimental writings and letters. He refused to discuss the philosophical implications of a unified theory, because to do so would detract from God's purposes which must be accepted on faith. He rejected those mathematical discoveries which he could not understand. His writing abounds with self-consoling rationalizations about the worth of the practical experimenter as against that of the theoretical physicist.

It is to Michael Faraday that this book is dedicated: in recognition of his greatness, in sadness at the fundamental tragedy of his life. Let there be no more humble men!

Introduction

Biological and social evolution have both demonstrated a fundamentally nonlinear process of universal development that can be appropriately characterized by a continually higher-order throughput of energy density within the biosphere. To enunciate the Second Law of Thermodynamics is to deny it: by the very act of speech. Man's history, our existence is proof of the negentropic process of universal development — at least to the full extent of man's reach so far.

But such a process of development is not preordained. Every past period of development within the biosphere has been characterized by crisis. Man's history has been shaped by his increasingly self-conscious efforts to solve these recurring crises. Briefly, the very success of any given society or animal species to reproduce itself within any given mode ensures that the resources on which such reproduction depends will be used up, creating an apparent energy crisis. This appears as a paradox only when energy is defined in terms of a given array of apparently fixed resources, rather than the increasing potentiation for man through science to discover and create new such resources.

Today, the frontiers of science lay in our emerging ability to understand and control energy dense plasmas in order to place at our disposal the incredibly enhanced energy potentiation implicit in a nuclear fusion-based technology.

Energy is a predicate of the self-development of the universe. We who have emerged from that process of development may now willfully order it to the deliberate advantage of our species. This is man's freedom. Just as there are no fixed limits to the resources at our disposal, so there can be no fixed laws which describe the universe.

Our present technology is still broadly based upon those "laws" of electromagnetism codified by James Clerk Maxwell in the nineteenth century. Twentieth century quantum mechanics merely formalizes the paradoxical basis upon which Maxwell rested his theory, which allows of treating reality only, alternately, as composed of particles or fields. Appropriately, it is within the domain of plasma physics that Maxwell's "laws" so far appear to be definitively violated.

In high energy dense plasmas, vortical phenomena occur which both violate the Second Law of Thermodynamics by increasing the energy gradient within the field, and, more specifically, violate Maxwell's laws by creating ion currents which capture and concentrate their own and surrounding magnetic fields. In particular, Lorentz's force law, which elaborates Maxwell's equations, is violated by these force-free vortices.

The "anomolies" which occur as nonlinear effects can presently be described only phenomenologically. High energy vortical formations are produced in situations where "classical" physics predicts the entropic dissipation of energy.

We are at the point where "classic" field theory must be superseded by a new physical-geometry which will recast the main body of experimental physics established in the nineteenth century.

In the process of developing a fusion-based technology, we will change the laws of the universe. The dreams of alchemists will seem merely pallid reflections of a reality which will place the willful creation of chemical elements at our disposal. Our understanding of this new lawfulness of which we are a determining part will expand accordingly.

The purpose of this book will be to explore the area of physical research, more or less demarcated by the last century, from the vantage point of that necessary task of subsumption of that work.The basis on which to approach the necessary new physical-geometry already exists in the work of Georg Friedrich Bernhard Riemann. During his short lifespan, from 1826 to 1866, he anticipated James Clerk Maxwell and, on all but secondary issues, Albert Einstein. Yet his work has been systematically distorted and obscured by comparison to Maxwell, H.A. Lorentz and Einstein for reasons which we shall discuss.

The same crisis which presents itself in the realm of theory is practically before us in the urgent necessity which we face to jump to a nuclear fusion power-based technology. Immediately, this demands the commitment of a large portion of global resources to the development of an appropriate technological infrastructure. The major resource, a highly skilled global labor-power pool, now exists only in embryo.

This is not an unusual crisis point in the history of the biosphere. Evolution has always proceeded through apparent discontinuities to achieve higher-ordered processes. The difference lies only in the degree to which man's willful intervention is now a determining feature of that process. Each previous stage of evolution has been delimited by the failure of then-existing energy resources. A period of linear growth and expansion of the biosphere would apparently occur, during which known resources were cannibalized, only to reveal an underlying nonlinear reality in which new resources discovered themselves. We term that tendency for development "negentropy."

It is no doubt the case that we are fortunate that dinosaurs could not decide on their own obsolescence, but the human species has always been afflicted with its own dinosaur faction which willfully seeks to hold back progress and thus condemn the masses of the world's population to plague, famine, and war out of their narrowly conceived self-interest to guarantee their share of shrinking, historically conceived fixed resources.

To maintain power, these humanoids must perpetrate the fraud that there is no longer hope for progress. They must convince large masses of the population foredoomed by their policies that there are no credible alternatives in the potential for an expanded higher-order technology on a newly developed resource base. Although particular philosophers may be unaware of the connection, reductionist empiricist philosophy has always served such masters by describing reality in terms of relations between fixed objects, and denying the reality of the process of development which subsumes these objects.

From the period of World War I on, large sections of the scientific community, centered in particular, around the Copenhagen School of Niels Bohr, were organized by Anglo-American Fabian intelligence services to the end of deliberately discrediting science in the eyes of the broader population. Bertrand Russell was one of the major controllers of this intelligence operation. An extreme empiricist of the positivist-nominalist school, his major theoretical writings were directed toward denying the reality of universals. Only the indescribable particular exists. It can be "named," but it can not be known. Scientific statements about universals are, to Russell, merely value judgements to be discarded at will with all other elements of morality. Thus, man is reduced to passivity. He can name the universe, but he must never aspire to determine the course of its development. From this it follows that any notion of progress is a mere chimera. Russell combined his activities as a philosopher and "popularizer" of science with open political organizing as a so-called pacifist. His work in the ban-the-bomb movement had the same general intended effect as efforts by Naderites today to discredit nuclear power as an energy resource.

It will be useful to invoke Russell's own words.

Our quotation comes from an article written in 1924, issued by the Philosophical Library in 1959, and entitled *The Future of Science:*

> Science has increased man's control over nature, and might therefore be supposed likely to increase his happiness and well-being. This would be the case if men were rational, but in fact they're bundles of passions and instincts . . .
>
> Scientists invent continually more elaborate methods of attack and defense. The net result of their labors is to diminish the proportion of the population that can be put into the fighting line,

> since more are required for munitions. This might seem a boon but, in fact, war is nowadays primarily against the civilian population....

This cynicism about the aims of science is accompanied by ridicule of its methods. We shall later discuss the theory of relativity at greater length, but at issue here is Russell's blatant use of that theory to debunk the notion of progress. The next quotation is from his 1925 work, *The ABC's of Relativity:*

> The collapse of the notion of one all-embracing time, in which all events throughout the universe can be dated, must, in the long run, affect our views as to cause and effect, evolution, and many other matters. For instance, the question whether, on the whole, there is progress in the universe, may depend upon our choice of a measure of time. If we choose one of a number of equally good clocks, we may find that the universe is progressing as fast as the most optimistic American thinks it is; if we choose another equally good clock, we may find that the universe is going from bad to worse as fast as the most melancholy Slav could imagine. Thus optimism and pessimism are neither true nor false but depend upon the choice of clocks.

Such was the view of the relatively pacifist philosopher who suggested a nuclear first strike against those melancholy Slavs at the close of the Second World War.

This kind of misuse of science by an accredited academic philosopher who did know better marks Russell as a deliberate fascist. Russell and his coterie at Trinity College, Cambridge were equally influential in the development of the positivist Copenhagen School ideology which surrounds atomic physics with the miasma of the uncertainty principle and acausality. Since both the momentum and the position of a particle cannot be precisely measured at the same instant, they claim that it follows that the universe is fundamentally unknowable and, therefore, potentially irrational.

The theory of relativity is treated by these ideologues as the bounding point of "classical" physics. By this they mean that Einstein extended Maxwell's field theory to the farthest possible bounds of scientific rationalism. In place of the commitment to a monistic view of the universe which informed all of Einstein's efforts, they have sanctified the dualism between the particle and field which flawed his and Maxwell's work. The Copenhagen School simply embellishes traditional British empiricism with a new title to the realm of complementarity where mind and body, subject and object, particle and field reign forever separate, but coequal.

Positivism did not start in the twentieth century. British empiricism has its roots in the failure of the sixteenth century Tudors to finally overcome the bestiality epitomized by the combination of Fugger banking and a Hapsburg-oriented feudal caste. Francis Bacon, political mentor of the abortive Essex counterrevolution and renowned father of the British empiricist school, united in his person just such a disgusting combination. The failure of the succeeding Cromwell commonwealth brought the period of the English renaissance to an end and created the climate in which Newton and Locke, a half-century later, could pick up the thread of Bacon's empiricism.

The political history of empiricism has never been praiseworthy, but its rampant fascist implications are products of this century. Nineteenth century scientists like Maxwell and Heinrich Hertz were avowed empiricists, yet their morality is not in question. Maxwell's commitment to progress shines through the pages of his popular works on science. His basic ideas may have been muddled, but his pedagogy is clear and direct in contrast to those modern writers who appear to amuse themselves by confusing the reader. "You see, you really are too stupid to understand science. Leave it to your betters to decide on those policy questions which are fundamental to the future of the human species." No such sneers from Maxwell! Nonetheless, the basic positivist-empiricist flaws in Maxwell and Hertz's work left them open to exploitation by men like Russell for whom they would have had only contempt.

These flawed conceptions must be addressed. They represent a detour from the achievements of Riemann which we can no longer afford. The view that man's knowledge is limited to the prediction of the probable lawful recurrence of a fixed succession of events, by denying the critical necessity of man's creative intervention in the universe, would condemn our species to virtual extinction in the next century.

Riemann's solution to the paradoxes of "modern" physics is the notion of a self-developing universe. Man's thought processes are at once a part of that universe and a reflection upon it. Since the body of man's scientific practice is constantly expanding, such a process of reflection must be broadly appropriate to the actual lawfulness of the universe. Furthermore, man is capable of directly reflecting upon his own scientific practice. This ability of man to form a conception of his own conceptual process and then to reconceptualize the appropriateness of these notions is itself a higher-order geometry which exists within the conceptual domain of the mind and thought, and is, at the same time, reflected in the neurological mapping of this thought to the brain.

This was the approach which shaped Riemann's development of a geometry of ordered manifolds. Each given description of reality represents a manifold which can approximate reality by simple linear extension, but which in turn must be superseded by a higher-order manifold which reflects upon the process of its creation and supersession. This is the epistemological standpoint we shall develop and return to throughout this book. We shall quote from his treatment of physical-geometry in order to establish a standard by which to judge the epistemology of Maxwell and Hertz.

What follows is from *On the Hypotheses Which Lie at the Foundations of Geometry* by Riemann:

> Determinate parts of a manifold, distinquished by a mark or by a boundary, are called quanta. Their comparison as to quantity comes in discrete magnitudes by counting, in continuous magnitude by measurement. Measuring consists in superposition of the magnitudes to be compared; for measurement, there is requisite some means of carrying forward one magnitude as a measure for the other. In default of this, one can compare two magnitudes only when one is a part of the other, and even then one can only decide upon the question of more and less, not upon the question of how many. The investigations which can be set on foot about them in this case form a general part of the doctrine of quantity independent of metric determinations, where magnitudes are thought of not as existing independent of position and not as expressible by a unity, but only as regions in a manifold....
>
> In a concept whose various modes of determination form a continuous manifold, if one passes in a definite way from one mode of determination to another, the modes of determination which are traversed constitute a simply extended manifold and its essential mark is this, that in it a continuous progress is possible from any point in only two directions, forward or backward. If now one forms the thought of this manifold again passing off into another entirely different, here again in a definite way, that is, in such a way that every point goes over into a definite point of the other, then all modes of determination thus obtained will form a doubly extended manifold. In a similar manner, one obtains a triply extended manifold when one represents to oneself that a double extension passes over in a definite way into one entirely different, and it is easy to see how one can prolong this construction indefinitely. If one considers his object of thought as variable instead of regarding the concept as determinable, then this construction can be characterized as a synthesis of a variability of n+1 dimensions out of a variability of n dimensions and a variability of one dimension.

In Riemann's conception, the particular point-object is constantly reevaluated as a subsumed feature of the self-expanding creation of higher-ordered conceptual manifolds. There can be no fixed position or momentum in such a universe. His answer to the Copenhagen School would be a hearty laugh that they thought they had uncovered a paradox separate from their own reductionist thinking practices.

Unfortunate Maxwell, in contrast, stocked the arsenal for Russell, the would-be destroyer of all those values which Maxwell treasured. In his introduction tc the book *Matter and Motion,* Maxwell writes:

> The name of physical science, however, is often applied in a more or less restricted manner to those branches of science in which the phenomena considered are of the simplest and most abstract kind, excluding the consideration of the more complex phenomena, such as those observed in living beings.
>
> The simplest case of all is that in which an event or phenomenon can be described as a change in the arrangement of certain bodies. Thus the motion of the moon may be described by stating the changes in her position relative to the earth in the order in which they follow one another.
>
> In other cases we may know that some change of arrangement has taken place, but we may not be able to ascertain what that change is.

Hertz elaborates the same premises. Reality is based upon the ordering of collections of things. Mental processes are fundamentally separate from physical processes. Man does not intervene to change the laws of the universe by his practice, he can only anticipate their unfolding. We quote from his *Principles of Mechanics:*

> The most direct, and in a sense the most important, problem which our conscious knowledge of nature should enable us to solve is the anticipation of future events, so that we may arrange our present affairs in accordance with such anticipation. As a basis for the solution of this problem we always make use of our knowledge of events which have already occurred, obtained by chance observation or by prearranged experiment. In endeavouring thus to draw inferences as to the future from the past, we always adopt the following process. We form for ourselves images of symbols of external objects; and the form which we give them is such that the necessary consequence of the images in thought are always the images of the necessary consequence in nature of the things pictured. In order that this requirement may be satisfied, there

> must be a certain conformity between nature and our thought. Experience teaches us that the requirement can be satisfied, and hence that such a conformity does in fact exist. When, from our accumulated previous experience, we have once succeeded in deducing images of the desired nature, we can then in a short time develop by means of them, as by means of models, the consequences which in the external world can only arise in a comparatively long time, or as the result of our own interposition. We are thus enabled to be in advance of the facts, and to decide as to present affairs in accordance with the insight so obtained. The images which we here speak of are our conceptions of things. With the things themselves, they are in conformity in one important respect, namely, in satisfying the above-mentioned requirement. For our purpose it is not necessary that they should be in conformity with the things in any other respect whatever. As a matter of fact, we do not know, nor have we any means of knowing, whether our conceptions of things are in conformity with them in any other than this *one* fundamental respect.

It is but a short step to the eclectisicm of twentieth century positivist Thomas S. Kuhn, a self-styled historian of science who peddles the view that theory is merely a convenient paradigm whose standard of acceptability is its ability to gain acceptance for its author. The following is his contribution to the young student who hopes to be a successful scientist: "does it really help to imagine that there is some one full, objective, true account of nature....?" — if you want your project funded buddy!

Aside from its moral implications, empiricism is inherently flawed, as Russell himself well knew. In his more honest, younger period, he collaborated with Alfred North Whitehead to write the *Principia Mathematica,* where they developed these implications. No given set of objects can include within itself the rules by which it was selected. Therefore, that rule must exist separately as a higher-ordered phenomenon. There cannot be a set of all sets. An amusing example developed by the mathematician Richards illustrates the point.

Take a list of numbers from one to ten and make another list of terms which are used to describe numbers. This list might include terms such as odd, even, negative, and fraction. Choose nine of them and add one further description known as Richardian, to be explained shortly. Mix the list of terms up and pair them randomly with the numbers as in the following example:

one	positive
two	odd
three	fraction

four	even
....	....
nine	Richardian
ten	multiple of three

As the number one is positive, the pair selection is appropriate and we shall move on. The number two, however, is not an odd number, therefore we shall put the number aside in a special set. Similarly, three is not a fraction and ten is not a multiple of three. We have now established a rule by which we are developing a new set of numbers in which set are contained the numbers two, three, and ten. Let us name this set Richardian numbers. Does the number nine belong in the set? If it does it doesn't, and if it doesn't it does!

The Richardian number is simply a parable to describe the set of all sets which do not contain themselves. Does this set contain itself? In this form, the question was discussed in the *Principia.* Georg Cantor gave a positive solution to the paradox by describing Riemann's continuing process of the creation of higher-order manifolds as the transfinite. The transfinite process of the creation of sets compares to the "bad" infinity of mere unending linear extension and the "truly" infinite conception which subsumes such extension within a given manifold.

The earliest treatment of the inability of a summation of particulars to be a description of a whole is ascribed to Zeno who asserted it in a geometrical form: the line is more than an infinite collection of points. He posed the problem that if a hare can travel twice as fast as a tortoise, he can still never overtake him, if the tortoise is given a head start. [Figure 1] Suppose that the hare waits for an hour to begin. After a further half-hour, the hare will have progressed to the point where the tortoise was, but the tortoise by then will have advanced.

Any progression from the part to the whole, the particular to the universal, necessarily introduces vicious fallacies. Any set of self-evident rules which appears to describe the behavior of a collection of objects necessarily abstracts from the process of development which accounts for the existence of these objects. Such rules can only be useful if they are explicitly acknowledged as linear approximations, properly subsumed within that development, and valid only within a certain delimited area.

A look at the simple charged particle underlines the ontological paradox. How is its charge measured? In order to merely determine if it is negatively or positively charged, it must be brought in the vicinity of other charges whose positive or negative charge has been previously determined, which in turn..., and so forth. Well then, there is no fundamental charged particle, but can we speak of a collection of particles? Place those charged particles together and, unless they are otherwise constrained,

Figure 1

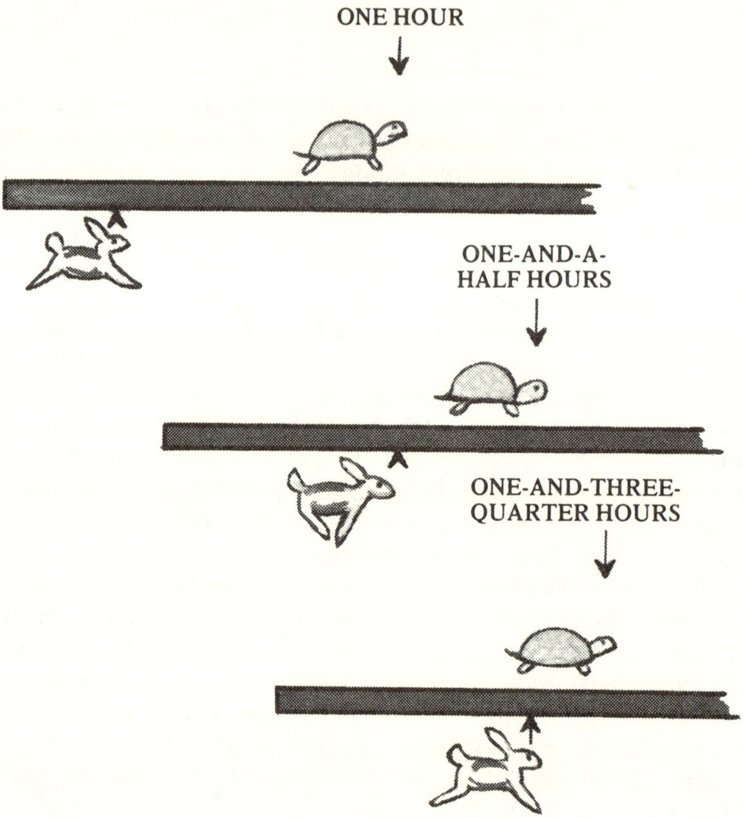

The distance between the hare and the tortoise will decrease, but they will always be separated by an "infinitesimal distance" at any point before the end of the line.

---

they will be set into motion until they can achieve an equilibrium state. They will no longer be simply a collection of charged particles because, as long as they are in motion, they will produce a magnetic field as a by-product of their motion.

Today the subject of electromagnetism is taught as a collection of basically incoherent rules. In our treatment, we approach the subject from the point of view of the electrostatic field in which charged particles are at rest. This standpoint is not inappropriate so long as it is the field which is the subject under which particle behavior is subsumed. The subject of electrostatics appropriately introduces the notion of a potential energy "function" which can describe the behavior of a collectivity of particles in terms of the underlying geometry which defines the given particles as a collectivity. We will be developing this notion of an energy function throughout this book; thus it is useful, at this point, to define the notion of function. A mathematical function is a rule which assigns a unique value to every particular position in a field, according to its position (or position and velocity, acceleration, etc.) in the given field. An energy function, on the other hand, characterizes the field by its energy throughput.

The subject of electrostatics as it is taught today, however, is premised on a false axiomatic basis which treats the particle as primary and the potential energy function as a convenient fiction. One starts with the behavior of charged particles which are constrained from moving, and then adduces the existence of electromagnetism from the behavior of a so-called isolated charged particle in an extraneously introduced magnetic field. In electrostatics, two positively charged electric particles will repel each other in the direction of the line which connects them. Similarly, two negatively charged particles will repel each other, while a positive and a negative charge will be drawn together. (The force will be directly proportional to the product of their respective charges and inversely proportional to the square of the distance between them — Coulomb's Law.) In electromagnetics, a charged particle in a magnetic field will experience a force directly proportional to the product of its charge, its velocity, and the magnetic intensity of the field through which it is traveling. But a charged particle, at its immediate point of intersection with the field, will be deflected in a direction which is, at the same time, perpendicular to its original direction of travel and the direction of the magnetic force field. [Figure 2] Why does this happen?

Electromagnetism implies a higher-ordered energy potentiation whose laws must subsume those of so-called electrostatics. In abstraction, Riemann provides a coherent approach; in particular, we must follow new lines opened by research in high energy plasmas in order to find an adequate reconceptualization of electromagnetic fields. The present critical treatment of classical field theory is only a step in that direction.

Any formalism which treats physics as little more than a collection of mathematical formulae has about as much connection to its own experimental roots as a pregnant Victorian lady had to the act of conception. The connection is embarrassingly vague. The opposite extreme is equally painful. Every serious natural philosopher assumed the connection between light, electrical, magnetic, and heat phenomena centuries before this could be empirically demonstrated or precisely determined.

Figure 2

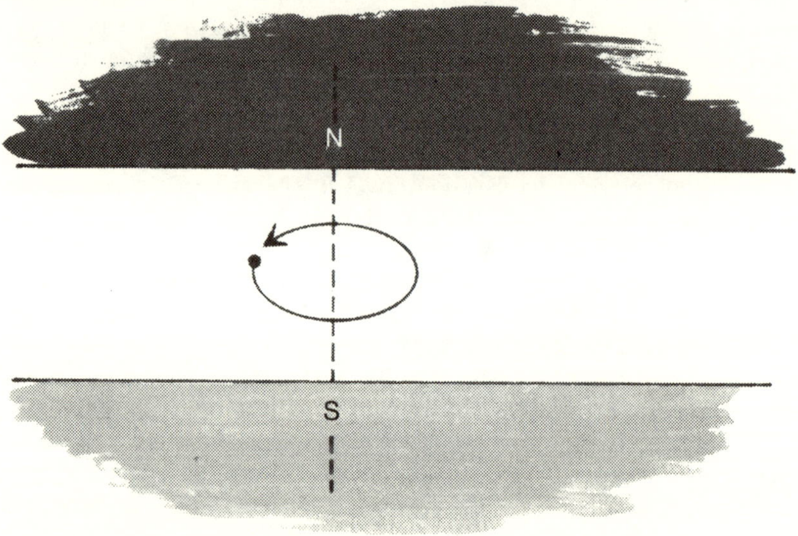

An isolated charged particle in motion can be deflected onto a circular path in a uniform strength magnetic field. The magnetic field extends from one pole to the other, while the plane of the particle's path is parallel to the poles' faces.

The actual process of experimentation was pitifully slow. The wave nature of light was denied by Newton despite conclusive evidence to the contrary. This error was not shared by his contemporaries Christiaan Huygens or Robert Hooke. (Newton's theory of light "fits" does not in any way even anticipate the particle-wave duality.) Only by a happy accident did Luigi Galvani anticipate in 1780 the action of a battery by identifying that the muscle of a frog was convulsed by an electrical discharge.

It was well known that lightning would magnetize the steel which it struck, but even after the production of batteries Hans Christian Oersted had difficulty in demonstrating the magnetic effect of a current in 1820 because of two mistaken presumptions. He mistakenly associated the production of magnetic effects with the heat produced by the resistance of a wire to flowing current. Thus, to his disadvantage, he chose a thin wire for his experiment and minimized the quantity of current. Secondly, he did not expect that a magnetic needle placed over a wire would be moved in a direction perpendicular to the wire, but rather thought that a wire laid in the perpendicular direction would be aligned with the wire. [Figure 3]

Figure 3

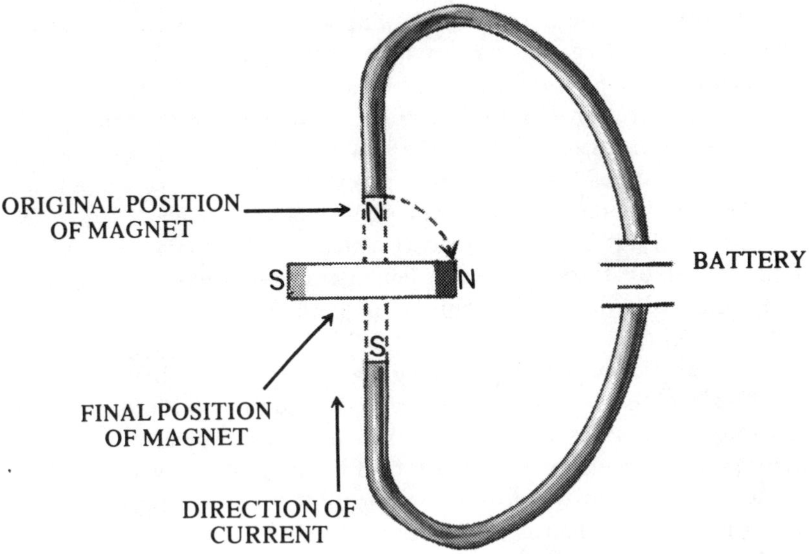

A bar magnet will align itself so that the line connecting the poles is perpendicular to the direction that current travels in the wire. Oersted thought that the magnet would be aligned to the dotted position in the figure.

Every serious scientist seeks those conceptions which unify apparently disparate phenomena — only the Fabian-controlled Copenhagen School could make ignorance a virtue. Even reductionists such as Maxwell and Hertz were concerned to prove that light is a form of electromagnetic phenomenon. Their problem was that they mislocated these unifying concepts as mere logical descriptions of ordered discrete particles, discrete events, discrete facts. To them, process was not fundamental, but a mere epiphenomenon; reality was fixed in the particular.

It is as much a pathetic error to suppose that science is best mastered by simply replicating its historical development as to suppose that it can be learned as a collation of formal rules. As Hegel pointed out in his introduction to *The Phenomenology of Mind*, each succeeding generation is able to recapitulate the achievements of the past in compressed form because it has absorbed these achievements as part of its own culture. The student already intuits those achievements which preceeding generations struggled to achieve; they are the artifacts of his culture. The number of cars, washing machines and the like per family in the United States and the fact that this country has the most highly skilled work force in the world is not a fortuitous correlation. The child of a typical American worker enters kindergarten with a high degree of technological competence gained in the garage and at the fuse box.

The correct approach to studying field theory today demands a certain multifariousness. Without the new foundation for mathematical physics which has yet to be elaborated except in bare outline, there is no way to completely circumvent the reductionism of modern texts. It is within the rigor laid out between their covers that technical competence still lies; however, these texts should only be approached from that standpoint already achieved within the discipline of mathematical physics by Riemann and Cantor.

We are suggesting a unique forcing approach to the problem of pedagogy. This book is divided into two sections. The first will introduce the underlying conceptions of potential theory, which represent both the vantage point from which to approach field theory as it is now understood and the ground upon which any fundamental breakthrough in theory must be built. Our intent is to recover the theoretical-experimental content of field theory which has been virtually obscured in modern treatments of the subject. Most of the mathematics which Maxwell synthesized in his theory was developed directly from the work of the continental Franco-German school. Today it is learned simply as a collection of mathematical formalisms.

It is desirable for the student to have the opportunity to replicate the experiments described in this book for himself in the laboratory, rather than to merely read of them or witness them in demonstration. If nothing else, the student is made forcibly aware of the absurdity of Newton's much quoted: *Hypothesis non fingo est.* Without a guiding hypothesis which is located in a broader theoretical overview which convinces the experimenter that the hypothesis must be broadly true, no experimenter except by lucky accident could plough his way through the welter of confusion which surrounds an experiment. The essence of successful experimental technique lies in weeding out the unessential deviations and discrepancies which obscure the point at issue, while recognizing those discrepancies which require fundamental readjustment of theory.

The horrendous lie of permissive Deweyite deschooling makes the point. "Put the naive child in a laboratory and let him decide what he wishes to do. Forbid the teacher to go beyond answering the child's question." Such an "anti-authoritarian" classroom can only end in chaos as the justly enraged child flings his microscope at the cowering teacher.

In any case, direct recourse to experiment is an appropriate antidote to formalism for the reason that no set of rules nor mechanical model can fully describe reality. The typical classroom experimental demonstration is weak not only because it builds from a premise of the falsely simple, but because it is designed to illustrate a narrowly delimited point. Variation from the expected result is considered to be a failure of the experiment rather than a prod to further investigation. The same criterion is applied to the student in a laboratory course where he is expected to "produce results" for a grade. This malady also afflicts professional laboratories funded on a delimited project basis. Needless to say, implied is the necessary large-scale funding of adult science education.

The second section includes excerpts from lecture notes on a year's course on electrodynamics given by Bernhard Riemann at the University of Göttingen, and prepared by one of his students after his death. The notes, of course, are only an indirect reflection of Riemann's thinking, but even so, they reveal the physical intuition which informed his more theoretical works as we shall show. Also included is Riemann's *A Contribution to Electrodynamics.* Through these writings, we establish the point made previously about Riemann's theoretical precedence. This section should challenge the average reader because to fully appreciate it demands an elementary knowledge of the calculus. It is our hope that the challenge to become mathematically literate will be accepted with joy.

In the next period, we will not be able to afford the luxury of a cheering section for progress. Even the most highly skilled workers will have to reeducate themselves to achieve the level of competence which will be demanded of them in a fusion-based economy. American know-how and inventiveness have been marveled at by travelers to this country since its inception. Only the ignorant and effete scorned the boisterous self-confidence which accompanied it, coining such terms as "Babbitry" to deride typical American boosterism. Even those immigrants who knew the most unremitting toil sacrificed so that their children could have an education and "get ahead."

Even more important, the present confluence of politics and science demands that every citizen have sufficient knowledge to intelligently inform his convictions. We must be competent to make those decisions upon which the future of the human race depends.

CHAPTER I

What Is Field Theory — The Potential

The apparent paradoxes which have arisen in quantum physics are the outcome of the failure of mathematical physics in the nineteenth century to develop a conceptual foundation for the appropriate notion of a self-developing universe. While Bernhard Riemann and Georg Cantor came closest to doing this conceptually, it is the formulations of James Clerk Maxwell which have survived to dominate practical accomplishments in electromagnetism. These formulations essentially overlook potential theory which provides their basis.

It is a commonplace myth that Michael Faraday was the first serious field theorist and that James Clerk Maxwell, his mathematical interpreter, was able to discover the mathematical electromagnetic radiation equations on the basis of this field theory.

The case is exactly the opposite. The mathematics which Maxwell used to develop Faraday's results came out of a body of work which had as its implicit subject unified field theory. Leonhard Euler, Pierre-Simon Laplace, Joseph-Louis Lagrange, and Karl Friedrich Gauss prepared these mathematical and theoretical foundations, elaborated by Sir William Hamilton, which shaped the positive content of Maxwell's work. Even the most seemingly abstract branches of nineteenth century mathematics were not disconnected from this project. Thus, Gauss's work in number theory allowed him to simplify astronomical calculations so that in three hours he accomplished what took Euler three days to complete at the cost of his failing eyesight. And Euler himself had astounded his contemporaries who previously needed three months to complete similar calculations.

As the story goes, Maxwell first elaborated the equations which describe the magnetic effects of an electrical current and the ability of a magnet in motion to induce electricity, and then, by algebraic substitution, came on the wave equations. In fact, James MacCullagh, a collaborator of

Sir William Hamilton, and Franz Neumann, a collaborator of Gauss, Wilhelm Weber and Bernhard Riemann, produced these same equations between the years 1839 and 1848, at least a decade before Maxwell began his scientific career.

Field theory, as it was developed through the work of Euler and Lagrange, elaborated by Gauss, and totally redefined by Riemann, depends upon the concept of potential energy. The term, potential energy, was first coined by J. Rankine at the end of the eighteenth century, but the notion of a work function followed directly from the work of Leibniz at the end of the seventeenth century. Euler and Lagrange's work in mechanics has been described, particularly by Anglo-American writers, as the rout of Cartesianism and the victory of Newtonianism in continental Europe. The truth is more complicated.

The practical and theoretical accomplishment of physics are interconnected, but not identical. Even on the basis of a hideous epistemology, it is possible for a practicing scientist to achieve limited gains by generalizing the existing technological practice of his society, and consolidating and linearly extending the work of his predecessors. Such was the case with Isaac Newton, who had the misfortune to live his adult life during the seventeenth century counterrevolution in England in association with such scum as John Locke.

"Force" vs. "Energy"

Newton synthesized the work of Galileo and Kepler, assimilating at the same time the ideas of such fertile contemporaries as Hooke, Huygens, and Boyle. His *Opticks* and *Principles of Natural Philosophy* in that sense are legitimate landmarks in the evolution of scientific thought. However, he used the legitimate correction of certain errors in Descartes's physics as a pretext to attack Cartesianism. This was part of a calculated factional attack on Renaissance traditions in England and on the continent which accompanied the physical elimination of followers of Oliver Cromwell. The extent to which Newton understood his role is perhaps moot. He was driven to direct political collaboration with Locke by his overpowering desire to escape the dreary confines of Cambridge University, and he ended his life as Keeper of the Mint under the reign of William of Orange. Newton and Locke were responsible for the policy of reversing the debasement of British coin to the profit of Dutch creditors.

Newton considered the inert atomized particle to be the fundamental invariant of the universe. He explicitly rejected the Cartesian notion of the primacy of the process of self-perfection. His rejection of progress placed him in the same camp as the Reverend Malthus and Herbert Spencer. (In Newton's day, the contention that the planet was overpopulated was gratuitous since the results of counterrevolution were already manifest in an epidemic of bubonic plague.) Newton, through his apologist in the controversy with Leibniz, Dr. Clarke, posited that the universe would constantly run down without the direct intervention of God. He attacked Descartes's notion of matter-in-motion on a theoretical as well as a practical basis. Descartes asserted the law of conservation of quantity-of-momentum as his approximation to the law of conservation of energy. Newton chose this as a major point of attack.

His law of universal gravitation asserted that every particle in the universe attracts every other with a force whose direction is that of a line joining the two, and whose magnitude is as the product of their masses and a constant, divided by the square of their distance apart. This force, however, is extraneous to the particles which are inelastic, hard balls. Two such balls coming together, if they were *the* fundamental particle, would come to absolute rest. Through this collision process, progressively larger amounts of energy would be removed from the universe, causing it to run down. On the basis of his delusory atomism, Newton rejected Leibniz's discovery of the conservation of energy. This conservation of the "living" motional energy of the universe, the product of mass and the square of the velocity, Leibniz called *vis viva* (which is double what we know as kinetic energy).

Epistemologically, the controversy between Leibniz and Newton over whether momentum or energy is fundamental is the same as the debate against Malthusianism. Leibniz enunciated the principle of conservation of energy and put it on a sound, though limited mathematical basis. (The full mathematical elaboration of the conservation of energy through conversion to heat, chemical, electrical, and other forms of energy awaited the nineteenth century and the work of Helmholtz, although it was already implied by Descartes.) Newton's universe was run by a clockmaker. It's animating principle was arbitrary. Its inherent limited lawfulness, the law of central forces, dooms it to inertion.

Mathematically, a force law can easily be resolved into a statement about energy (see below). Hence, the argument between the continental school of physics and the Newtonians was not really a mathematical-physical argument at all. What Newton and Locke did was to isolate a series of gravitational relationships, for which a force law provides a useful first approximation, and then reified this law to rationalize an attack on continental humanism.

Force is measured by a change in momentum, the acceleration or deceleration of a body. Since a body in motion will tend to retain that motion, it is the effect of force upon a body which changes its velocity. As velocity is only a relational concept, the law of inertia is really only a reflection of the fact that an isolated body does not have velocity.

Since for Descartes and Leibniz the invariant property of the universe was its capacity for self-development, it was inconceivable that the universe should be tending to come to a halt. Leibniz, rejecting the notion of dead matter, replaced the fundamental particle by animated monads which incorporated his notion of development. Under these conditions, motion could be conserved because it was embodied within the pores of the universe. Looking for the form of the conversion of this motion, he came upon the relationship between the velocity of a falling body and the work necessary to lift that body. This quantity of motion, which is preserved, double the kinetic energy, or mv^2, is always a positive scalar quantity. Momentum, on the other hand, is a directed vector quantity, and is either positive or negative. The difference is illustrated by the case of two cars which approach each other at a speed of 50 miles per hour. At the point of collision, apparent motion stops. In modern terminology, the macroscopic momentum was always zero, $(+mv)+(-mv)$; the energy however was mv^2, where m denotes mass and v denotes velocity. It is transformed into the explosive effects of collision. For Descartes, the distinction of before and after the crash — rest and motion — was critical.

It is worthwhile to note that both Descartes and Leibniz were led to a paradoxical formulation for the very reason that they attempted to quantify a universal invariant in terms of a summation of particulars, objects in motion. While Newton's universe squatted obscenely in some absolute space, and therefore was not the universe, Descartes and Leibniz recognized space and time, and therefore velocity, as relationships. If there is no absolute space, there is no absolute velocity or absolute rest.

Because they failed to conceptualize energy as the characteristic geometry of a field, but rather in terms of the additive sum of the action of its parts, they ran into a paradox.

How then is energy characterized in terms of the particular, if there is no absolute standard of velocity? Take two bodies of equal weight, one moving with a velocity of five miles per hour and one with a velocity of three. But why not simply assign them their relative velocity of two miles per hour. Clearly another object is involved which acts as a standard and is considered to be at rest. Consider the energy of this system of three bodies. Suppose now that we considered the object moving at three mph as zero and attribute a velocity of negative three miles per hour to the body previously considered to be at rest. The other two bodies will now be moving at a velocity of two and negative three miles per hour, respectively. Now compute the energy of the system! We have not contradicted the invariant quality of energy, provided we maintain a fixed standard of measure. The total quantity of energy in the universe is unchanging. We have established that any such fixed reference system is arbitrary, so that our measure of universal energy is equally arbitrary. By the principle of sufficient reason, we know that we have still only approximated a notion of the invariance of energy.

The mathematical resolution of the conflict between the Newtonians and the continental school is straightforward:

The energy-work relationship is expressed quite simply in mathematical terms. Force is measured as change in momentum or mass times acceleration. Work is the amount of force expended over a given distance. The work performed per unit of time is equated to the change in kinetic energy. Thus, work measures action. The capacity of the system to perform future work is known as its potential energy. In simple reproduction theories, this is equated to past work plus an unknown "hidden resources" constant. The total energy of a system is the sum of its kinetic and potential energy. This amount may or mat not be conserved although it is treated as a fixed quantity within the universe as a whole.

Momentum is denoted by mv.
Force equals change in momentum:

$$F = \frac{m(v_2 - v_1)}{t} .$$

Displacement is denoted by s.
Work is denoted by Fs.
The mean velocity over a given time interval is

$$\frac{v_2 + v_1}{2} = \frac{s}{t} .$$

Hence,

$$Fs = \frac{ms(v_2 - v_1)}{t} = m\left(\frac{v_2 + v_1}{2}\right)(v_2 - v_1) = \tfrac{1}{2}mv_2^2 - \tfrac{1}{2}mv_1^2 .$$

Work per unit time is the change in kinetic energy.

Leibniz accompanied his discovery of the indestructibility of energy with a theological argument upon which he grounded his debate with Newton. If God is omnipotent and omniscient, then he could not have created an imperfect universe. A universe which demands God's intervention lest it become inert would lack perfection. Being omniscient, he would have foreseen and remedied such imperfection in advance of creation. If God is omnipotent as well as omniscient, then he must still retain the possibility of intervening in the universe. As his intervention can not be to correct defects, it must be to the end of creating successively higher orders of perfection. Such was the idea which Voltaire ignominiously satirized in his parody *Candide*, where the wretched Dr. Pangloss goes around muttering: "This is the best of all possible worlds" — in the face of every calamity.

On the basis of this notion of perfection, Leibniz established the principle of sufficient reason as the axiomatic grounding of all knowledge. No event can be properly explained merely by its sequence in a chain of preceding events. It must be lawful in terms of the ordering of the universe as a whole. An adequate theory must locate the particular event within the process of development of the whole.

The principle of least action was a metaphysical bridge for the French materialist school of natural science. The principle states that, in accomplishing a motion between two points, a particle will choose the path for which the action — the product of its momentum and the displacement over which it retains that momentum — will be the least possible amount. It can be described as a kind of accountant's principle which nonetheless assigns a primacy to process rather than to the particular. The theological version of the principle is an appropriate point of reference. God has ordered the universe in such a way as to maximize its efficiency of operation. Natural law is determined as that sequence of events directed toward a goal in which the necessary amount of action required is minimized. The principle of least action connected the notion of perfection to the notion of conservation of energy by degrading the notion of perfection to that of coherence. If the quantity of energy in the universe is fixed, since it is a self-subsisting system, and if there is to be perfection, then that quantity must be utilized with ever greater efficiency. Because they shared a notion of absolute perfectibility of the universe rather than relative perfection, neither Descartes nor Leibniz made use of the term, although Descartes popularized Snell's law of refraction, which can be effectively stated in terms of a principle of least action.

The extension of these least-action or minimizing principles led, in the eighteenth and nineteenth centuries, to the calculus of variations. Natural law was established as the description of paths of least action. The Cartesian natural law, which describes the evolving lawfulness of an expanding universe, was replaced by a body of fixed natural laws.

The Lagrangian Universe Supersedes Newton's

Euler, Lagrange, Laplace, and the school of French materialism (which must be extended to include that inveterate German nationalist Gauss) were not "Newtonians," although they applied the methods of analysis and, in particular, variational principles to the corpus of mechanics delineated by Newton. But, they were not Cartesians either. Laplace is perhaps most famous for giving expression to the metaphysical presumptions of that school of French materialism with his complacent assumption that a supranatural physicist apprised of the position and momentum of every particle in the physical universe could predict the entire course of future events on the basis of the laws of physics as they were then known. God was still omniscient, but he had lost his potency in a universe which was unalterably lawful.

Under such presumptions, perfection was limited to the realm of subjectivity. Scientists could hope to constantly penetrate the universe more deeply, Laplace notwithstanding, but they could not expect to fundamentally change the existing lawful ordering of natural events, nor could they expect an evolutionary natural, as opposed to social, process to be occurring independent of their intervention. This notion of a preexisting perfection, to which man can only approximate, was exactly the notion which Descartes rejected by going back explicitly to the great achievements of the Arab Renaissance and such thinkers as Ibn Sina rather than the later works of the Italian Renaissance philosopher Ficino.

Descartes located man's creativity, his ability to conceptualize on a higher and higher order of universality, as the direct correlative of the self-development of the universe. The human species' creative capacity is not only appropriate to understanding the ordering of natural events, it is also the appropriate culmination of that ordered process of evolution from which we have emerged. The standpoint of eighteenth and nineteenth century French materialists represented a metaphysical retreat from the heights of the seventeenth century, just as the French Revolution (not the American) represented a retreat from the Tudor industrial practice of Louis XIV's Finance Minister Jean-Baptiste Colbert and the English Commonwealth's Oliver Cromwell. But they were not Newtonians.

French materialists reconceptualized Newtonian mechanics in terms of fluid dynamics. Typically, textbooks are careful to point out that Lagrangian mechanics, for example, is interchangeable with Newtonian mechanics. They treat Lagrange's characterizations of whole systems of particles as self-contained wholes by using generalized phase-space coordinates as merely a convenience over Newton's original formulation. While this is true formally, it is such formalism which destroys thought. Lagrange, Euler, and so forth, whether or not they were fully aware of doing so, laid the mathematical basis for field theory precisely because they treated collectivities as primary in practice. The notion of a generalized phase space is developed by taking as coordinates of a system not the positions of the particles of which it is composed, but, instead, certain parameters which appropriately characterize the behavior of the system as a whole. One such simple characterization might be the center of mass, which characterizes the stability of an extended body, for example, a chair. (In Newtonian terms, such an extended body would be the epiphenomenon of a collection of atoms.) The velocity of such a system or its temperature (the average kinetic energy of its molecules) are equally characteristics of the system as a whole. The point is not that Newton did not in practice know that he sat on chairs nor that he failed to use such simple devices as center of mass for calculation, but the systematic reconceptualization of systems as such, particularly in the area of fluid dynamics.

The Lagrangian universe was not the self-destructing fundamentally anarchic universe of Isaac Newton. They took Newtonian particle mechanics and reconceived it in terms of metastable collectivities rather than collections of individual particles. A universe which is not rapidly devolving must maintain internal differentiation; while, at the same time conserving its energy, it must conserve its capacity for action. It must resist the entropic tendency to degenerate into a homogenous blah in order that the throughput or energy flow of the system can be maintained. In order to resist the entropic tendency toward loss of distinction and homogeneity, such a system must have an internal tendency to reproduce itself which is an essential characteristic of the system. The freedom of action of the "particles" which compose the system is delimited by "constraints," such as the material cohesive binding force, or the elasticity of a string on which the motion of a pendulum depends, and so forth. In such a system, the particle is determined by the whole. This is essentially a field theory. Its weakness is its failure to comprehend the possibility of development to a system with higher-order constraints. The term "constraints" is instructive. The system is constrained from entropic dissolution, but development is devolution.

This inborn flaw is also characteristic of and consistent with the notions of political economy which prevailed among the French Physiocrats during the same period. While the Physiocrats lacked a Colbertian-Tudor understanding of the necessity for industrial progress, they did study the cyclical process of reproduction in the real economy. By seeing a fixed mode of agriculture as the only source of real value in the economy, they fell into the same trap as the natural scientists. But, by contrast with ecomomists of the school of Locke, Hume, and Ricardo, who actively conspired to achieve the self-destruction of the universe by a policy of monetarist looting, they were humanists.

From the standpoint of simple reproduction, the universe is fundamentally stable and unchanging. Work is the activity necessary to restore the system to a given condition following any dynamic interaction which it may have with its environment. Such a notion expanded to the domain of political economy includes the differentiated notion of the maintenance of the labor force and the material preconditions for labor. It can even include a fixed surplus which is assigned to support the aged and infirm, a more or less fixed population of youth, or a ruling strata.

The essential idea is the same in Lagrangian physics and physiocratic political economy. Work performed on a system from outside the system increases its energy by an equivalent amount of "free" energy which defines its capacity to replicate that work. This capacity to perform work is its potential to release motional or kinetic energy.

The system may adopt a cyclical pattern of behavior, as is the case with a pendulum which alternately "consumes" and performs work when set into motion. It is able to absorb a certain rate of increase of free energy and maintain itself in an increasingly dynamic equilibrium. If such a pendulum is violently displaced by an excess of free energy, its material constraints may be broken. Stability will be established, but the pendulum will be destroyed. Energy will be conserved within the environment as a whole, but at the cost of the devolution of the subsystem. The universe will become less ordered.

At any given period, such a system will be most stable when its change in potential energy (its stored capacity to perform work, which includes its inertial or elastic energy) is balanced by its change in kinetic energy. This is a productive system which does not expend more energy than it has on hand to replenish its stocks, but does not accumulate an increasing store of stocks. Such a system may become more stable with an increase of velocity, as in the case of a bicycle wheel, but the balance between the potential energy and the kinetic energy is still maintained. The balance need not be zero, but the system will be in equilibrium when the balance is maintained at a minimum. It will be most stable as that difference approaches zero.

Jean LeRond d'Alembert, in 1743, reformulated Newton's third law of mechanics, "every action gives rise to a reaction," in terms of a system in equilibrium. If such a system is at rest, then clearly any forces exerted on the system are counterbalanced by the reaction of the system. I am able to sit on a chair without collapsing to the floor which collapses. . . which collapses.

In a system at rest, no work is done; nonetheless, we can speak of the virtual work as the tendency of the counterbalanced forces to do work. The sum of all of these tendencies, virtual displacements, must be zero. If the system is in motion, the problem can be transformed into an equivalent statical problem by the following consideration. Every system in motion tends to maintain that motion by the principle of inertia; this tendency can be considered as an inertial velocity potential, an added constraint on the system, which increases as a function of motion.

In a great advance over Newton, Euler and Lagrange developed the notion of generalized coordinates which describe a system not in terms of the enumeration of its particularities, but in terms of the motion of the system as a whole. The motion of a system can be considered as a combination of a translation plus a rotation. If such a system is considered in terms of quasi-static transformation, it will not only have inertial potential energy due to its translational motion, but it will have a rotatory potential. An observer standing on a rotating platform would experience this rotatory potential in two ways: as a centrifugal force tending radially outward from the axis of rotation, and as a Coriolis force, a force which would manifest itself only if the observer tried to walk across the platform and which is perpendicular to the observer's direction of travel and proportional to his speed. [Figure 4] Furthermore, by applying a torque to the platform, an angular acceleration can be induced.

The parallel with electromagnetic forces is suggestive. The Lorentz force law asserts that an electron moving through a magnetic field experiences a force similar to the Coriolis force in that it is directed perpendicular to the initial direction of travel and proportional to the velocity. The torque compares to induced electromotive force. An electrical current will be induced by variation of the magnetic field. The analogy is between the angular velocity of the platform and the field produced between a current and a magnet, or a current and another current.

A simple experiment which demonstrates the perpendicular force can be carried out by suspending a bicycle wheel from the ceiling by a rope attached to its axle. [Figure 5] Spin the wheel and it will also inscribe a circle around the rope. This added motion of the wheel is perpendicular to the gravitational force and to the angular velocity of the spinning wheel. Such an effect can be resolved algebraically to particle dynamics, but such algebra merely emphasizes the idiocy of reductionism.

The expression for the total energy of a system must be a function not only of its coordinates, but at least of the rate of change of these coordinates. The restriction to that rate of change or first derivative is, of course, only an approximation. The degree of approximation which is both desirable and acceptable is defined on the one hand by the limits of existing technology and, on the other, by the requisite precision.

The Lagrangian equation — the typical form of the variation energy equation — was used by Riemann in his work on electromagnetism, despite the obvious (to him) flaws of such physiocratic physics. The approximation of potential theory, implicit in this kind of simple-reproduction model, allowed him to test the coherence of his electron-field theory. The point at issue is to determine that the laws of conservation of energy are not violated and that, at the same time, the system is operating according to the least-action principle of efficiency which, by approximation governs systems which operate under "natural" law. Because the rate of evolution

Figure 4

A. Centrifugal force is experienced as a tendency to fly off the disc.

B. When moving toward the center of a rotating disc, the observer will feel the force pulling him sideways. This is the Coriolis force.

of physical systems is relatively low at middle-range energy values, such approximations are useful in the first instance. The limitations of Hamiltonian-Lagrangian mechanics at either end of the spectrum are an acute disadvantage today — in the case of superconduction or high energy dense plasmas.

The Lagrangian expresses the minimizing principle (principle of least action) with reference to the balance of potential and kinetic energy in

Figure 5

A bicycle wheel suspended from its axle and spun will act like a top. Instead of dropping so that the axle is vertical, the wheel precesses about the rope. This surprising phenomenon is usually explained in terms of the torque acting on the wheel due to the gravitational force. The torque is perpendicular to both the gravitational force and the axle. Hence, since a change in the angular momentum (directed along the axle) must occur in the same direction as the torque, the wheel will precess rather than fall.

terms of their dependence on both the generalized coordinates of a given system and the rate of change of these coordinates. If q represents any coordinate by which the system is described, $\dot{q}$ will denote the derivative or rate of change of that coordinate in time, where t denotes time. Let

L=T-V be the energy function where T represents kinetic and V potential energy; the Lagrangian function is then expressed as follows:

$$\frac{\partial L}{\partial q_i} - \frac{d}{dt}\frac{\partial L}{\partial \dot{q}_i} = 0.$$

On substitution, this expression takes the equivalent form:

$$\frac{\partial T}{\partial q_i} - \frac{d}{dt}\frac{\partial T}{\partial \dot{q}_i} = \frac{\partial V}{\partial q_i} - \frac{d}{dt}\frac{\partial V}{\partial \dot{q}_i}.$$

From this latter form of the equation, we can see that the change in the kinetic energy of an object moving from one point to another is identical to the potential energy difference between the two points. However, the functions of potential and kinetic energy are now expressed both in terms of position (q) and of velocity ($\dot{q}$). Furthermore, this equation includes Newton's force law as a special case, since $(\partial L/\partial q_i)$ defines the force in terms of potential energy.†

The study of such minimizing principles, known as the calculus of variations, was a major achievement of Euler, Lagrange, d'Alembert, Hamilton, and Dirichlet. However, one of the earliest such problems was considered by the ancient Greeks. The problem was to find, among all closed curves of a given length, the one which encloses the maximum area. It was intuitively evident that the solution was a circle, but this fact has been satisfactorily proved only in recent times. The modern systematic development of the calculus of variations began in 1696 with the formulation of the brachistochrone problem by John Bernoulli. [Figure 6] The problem was to find, among all curves connecting two given points, that one which has the property that a particle sliding along it, under the action of gravity alone, falls from one point to the other in the least time. Bernoulli solved the problem by analogy to the law of the refraction of light in different mediums.

It was Descartes who formulated the basic equations which connected the path followed by a light ray with the variation of the

† The derivative of any function, df(x)/dx, represents the rate of change of the function with respect to changes in the designated variable. Graphically, the derivative can be represented as the slope of the function when it is plotted against the variable. For functions of more than one variable, the partial derivative, $\partial f(x,y,z)/\partial x$, is a useful operation. It is used to determine the rate of change of the function with respect to the designated variable, while the other variables are held constant. In addition to the derivative, the other major operation of calculus is the integral of a function, $\int f(x)dx$. Graphically, this corresponds to calculating the area marked off between the curve representing the function and the independent variable axis. The operations of differentiation and integration are inverses of each other, so that the derivative of the integral of a function is the function itself.

Figure 6

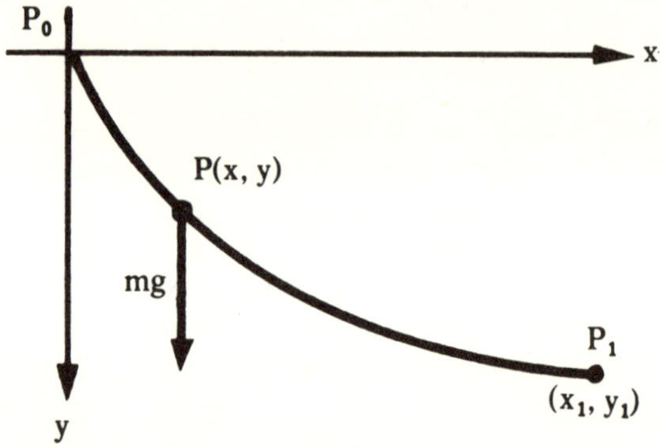

The brachistochrone problem can be stated as follows: "Among all smooth curves joining P_0 to P_1, find the one along which a bead might slide, subject only to the force of gravity, in the shortest time." At first sight, one might guess that the straight line joining the two points would yield the shortest time, but there may be some gain in time by having the particle fall vertically at first, building up velocity more quickly than if it were to slide along an inclined path. The solution to the problem is in fact an arc of a cycloid through P_0 and P_1 with a cusp at P_0, as shown in the figure.

---

medium through which it travels, on the basis of experimental work by Willebrod Snell. [Figure 7] He proposed the correct equation $u_1 \sin i = u_2 \sin r$, where u_1 refers to the refractive index of the first medium and u_2 to the second. However, Descartes assumed incorrectly that the speed of light in two different media varied directly as the refractive index of the media. In fact, the variation is inverse.

Descartes's argument can be stated as follows. Assume that your media are proportional as described, then

$$\frac{u_1}{u_2} = \frac{v_i}{v_r}.$$

Consider that the light travels an equal time before and after crossing the boundary to the two media. The horizontal velocity of the light must be the same in either medium since they share a common boundary. Call that velocity v_H. Draw a perpendicular through the point of intersection of the light with the boundary. Then, in the case of light,

Figure 7

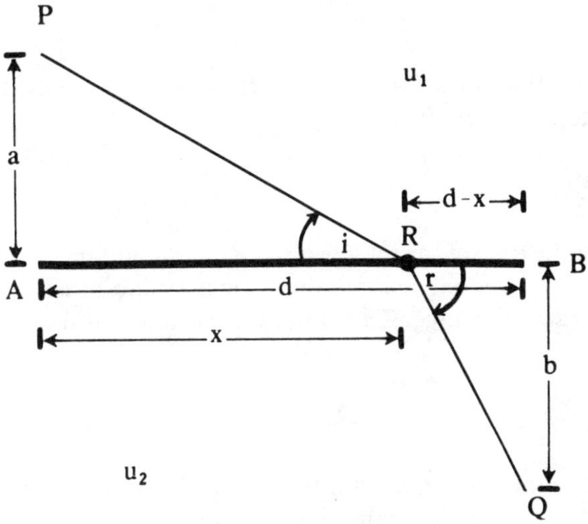

Before refraction, light travels from P to R at speed v_i in medium u_1; after refraction, it travels from R to Q with speed v_r in medium u_2. The line AB represents an optical interface, say between air and glass. In passing from air to glass, the path of the light is refracted so that the ray is more nearly perpendicular to the interface.

$$v_i \sin i = v_r \sin r = u_1 \sin i = u_2 \sin r = v_H ,$$

$$v_i = v_r, \ \measuredangle i = \measuredangle r.$$

Descartes, of course, lacked the means to actually measure the velocity.

Pierre de Fermat established the correct relationship between the velocity and refractive indices in 1661. The velocity decreases with the increase of the refractive index, it varies inversely. He also formulated the principle of least time as follows: "Nature always acts by the shortest course."

$$\frac{u_1}{u_2} = \frac{v_r}{v_i};$$

therefore, if t denotes time,

$$t = \frac{PR}{v_i} + \frac{RQ}{v_r}.$$

If a is the length of PA and b the length of BQ, then

$$t = \frac{\sqrt{a^2 + x^2}}{v_i} + \frac{\sqrt{b^2 + (d - x)^2}}{v_r}.$$

To find the value for x for which the time is a minimum, take a small variation of the path since

$$\frac{dt}{dx} = \frac{1}{v_i}(\sin i) - \frac{1}{v_r}(\sin r).$$

In order for the time of the total trip to be at a minimum, any small variation of the path can only vary the time by a second order difference, the rate of the rate of change. Therefore, the variation in time can be considered to be zero.

$$\frac{dt}{dx} = 0; \qquad \frac{\sin i}{v_i} = \frac{\sin r}{v_r}.$$

A simple example will be helpful to explain that equating the derivative, dt/dx, to zero gives a minimum. Consider the slope in a valley. At the very bottom there is no gradient, the valley is level. [Figure 8]

Bernoulli's solution to the brachistochrone problem was to consider that the vertical distance traveled by a particle was uniform in equal distances. He arrived at a solution by a limiting process where the vertical distances were decreased to zero.

Energy Throughput And Thermodynamics

The French materialist school of natural science took the collation of empirical mechanics by Newton and reconceptualized it in terms of the physiocratic world outlook which falsely characterizes the social reproductive process as circular rather than developmental or helical. Practically, this allowed the French materialists to simplify Newtonian mechanics by assigning coordinates to whole systems at once. Furthermore, having discovered that equilibrium tendencies could be systematically described by minimizing principles — the calculus of variations — they now asserted the axiomatic importance of these principles. Study of Newton's descriptive force laws revealed the underlying stabilizing tendencies as such. These tendencies could be described by Lagrangian

Figure 8

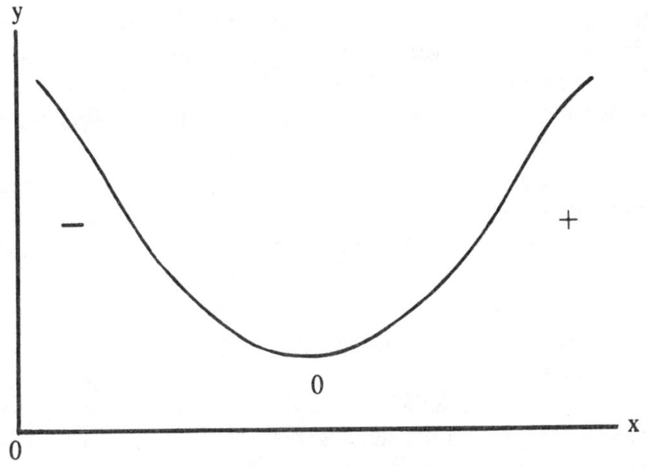

A simple example will explain how the solution to the equation which arises from equating the derivative of a function to zero finds the minimum value of the function. At the bottom of a valley, there is no gradient, it is level. On the left side, the slope is considered negative, on the right side, positive. At the bottom, the negative slope changes to positive, necessarily passing through zero. That is, the ratio of vertical to horizontal change, dy/dx, is negative on the left, positive on the right, and zero in the middle. By equating the slope (derivative) of the function to zero and solving for the independent variable, x, we arrive at that value of x for which the function is minimum.

---

energy equations. These equations could now be used as the basis from which to deduce the behavior of bodies operating under the influence of "natural laws" such as Newton's central force law which describes the gravitational interaction of bodies. This allowed a great simplification of calculations in practical applications in astronomy, engineering, and fluid dynamics and was the precondition for the development of field theory.

Philosophically, despite their retreat from the Cartesian notion of perfection, the work of the French materialist school was critical to our own ability to conceptualize the actual field theory appropriate to a self-developing universe which reproduces itself by lawful self-expansion rather than conservation. They located and emphasized energy throughput as the necessary universal invariant. It is only necessary to now conceptualize this invariance as a transvariance. By approximation, every particular period of universal development and specialized subsystems within a given period can be characterized and compared by the level of energy throughput typical for each. Each such period or subsystem has a characteristic energy density invariant. The transvariance of energy, then, characterizes the tendency of each succeeding period, or more developed subsystems within the same epoch to process a continually higher level of energy throughput both qualitatively and quantitatively.

Such a notion of energy throughput allows us the point of approach from which to escape the dilemma noted in the paradoxical description of energy as the summation of a function of the velocity of particular bodies in motion, $\Sigma m v_i^2$, even though velocity itself is only a relational concept. The same problem emerges in the law of conservation of energy, which should more properly be known as the law of conversion of energy. Coal is converted to electrical energy by traditional means at only a 40 percent level of efficiency. By such an accounting, when we burn coal to produce steam power in order to operate a generator, we are wasting natural energy. Clearly this is not the case. What is necessary is the obvious notion that an economy based upon electrical energy has a higher quantity of energy throughput than a coal-based economy. How can this be defined? Not on the basis of the transfer of kinetic energy between particles. By that standard, we agree that 60 percent of the kinetic energy of the coal and associated materials is dissipated into waste when we produce electrical energy.

Two related types of historical examples make the point for the natural sciences. (We are not stretching analogies — no adequate notion of energy could be established which ignored man's creation of higher energy forms through social processes.) The various social uses of electricity enormously increase human productivity as a whole. They lengthen the productive work day by lighting, establish global communication and therefore the expansion of the social division of labor, they allow a degree of controlled precision not previously possible. Because of this enhanced productive capability modern population potential is vastly increased over that of preelectric technologies. In practice we have established a standard of quality of energy throughput in the enhanced ability of human society to expand production qualitatively and quantitatively so as to sustain a greater population at a higher standard of living, and, at the same time, increase the potential for future such developments.

It is the nonlinear aspect of this qualitatively enhanced throughput of energy which provides our second group of examples. The ability to build batteries and therefore control a flow of electricity allowed chemists, such as Faraday's mentor, Sir Humphrey Davy, to use electrolysis to explore the chemical composition of matter. Thus, he was led to the discovery that chlorine is an element rather than a compound. This may be described, in general, as the ability of each society to redefine the potentiation of its resources. To a stone age man, iron ore was merely another kind of rock.

Before proceeding further a warning is in order. The theory of conservation of energy can become a mere sly semantic game. "Look we have discovered new energy resources; our supply of energy has expanded." "Ah, but they were there all along. You have discovered them, not created them."

A look back at Sir Humphrey and his pupil Michael Faraday breaks the impasse. The ability to build an electromagnet by passing a current of electricity through a tightly wound coil of wire emphasizes the point. New collectivities, new subsystems develop qualitatively new dynamic parameters. The torque of a screw or a bicycle wheel is not inherent in each separate particle of the screw or wheel as it rotates, but in the interacting collectivity of particles. The qualities of a collectivity are not the linear sum of its particularities. These collectivities demand new "force laws" which define their mutual interactions and the establishment of new constraints and boundary values which determine the conditions for their dynamic equilibrium with the environment as a whole. But no such event is merely local. A new universal potential energy gradient has been established, thereby, in turn, redefining the universal energy function and the particularities subsumed by it.

It is in just this area, the identification of the potentiality for new collectivities, that fundamental scientific breakthroughs occur. It is in the realization of these potentialities that progress occurs. It is the unique ability of the human species to willfully choose among alternate manifolds of such potentialities which determines the unique dynamic parameter of our species as a collectivity. It is this ability to willfully create new higher order universes which is our human nature.

It is the willful enhancement of that potential for creativity which provides the appropriate criteria for the direction of necessary social progress.

At the same time as energy theory was being developed in the nineteenth century and the existing energy conversion relationships were codified as the First Law of Thermodynamics, the law of conservation of energy, a parallel line of development led to the formulation of the Second Law, the law of the tendency for increase of entropy. The precondition for the development of the Second Law of Thermodynamics was the work of Sadi Carnot and others in the 1820s. These men were practically concerned to measure the efficiency of steam engines. Their concern was to maximize the throughput of energy.

Their work was taken up, in particular, by Maxwell, Rudolf

Clausius, and Ludwig Boltzmann. In the earlier part of the century, heat was viewed as a qualitative phenomenon. The term caloric was used to describe its substance-like character. Maxwell, Clausius, and Boltzmann, however, advocated the kinetic heat theory which only reemerged as hegemonic in the twentieth century. On the basis of a kinetic theory of heat, they refined Carnot's work and enunciated the Second Law.

The kinetic theory describes heat energy as the summation of the individual kinetic energies of the constituent molecules of a substance. This summation is perceived as heat rather than as motion because of the randomly directed motion of the interacting particles. They tend to oscillate and set a collective vibratory motion rather than a displacement in space. The molecular interactions involved in changes of state — from liquid to solid to gas — are only of quantitative interest to this point of view. Thermodynamics is statistical mechanics. Just those features of the behavior of a collectivity which determine energy transvariance are "averaged out" of consideration. Such a delimitation of focus can be useful within narrow limits. It can also serve as the basis for vicious rationalizations of the denial of the inherent tendency of the universe for self-development.

The Second Law of Thermodynamics is based upon the empirical evidence that heat "flows" in the direction of a decreasing temperature gradient — from hot to cold. It can only be transferred from cold to hot if it is generated as a byproduct of work, for example, by friction. The tendency of frictional heat is to escape a given system and to flow randomly into any neighboring environment, thereby reducing the temperature gradient without accomplishing work. It is this tendency for energy to dissipate or degrade itself, thereby flattening the potential gradient, which is known as entropy. It is the tendency for randomness, disorder, and homogeneity; the chaos of anarchy, rather than freedom within ordered necessity.

A schematic look at the functioning of a steam engine illustrates the conceptions first developed by Carnot. Let temperature T_2 be greater than temperature T_1. Let there be a boiler at position A. Place a piston containing water on position A. The water will be heated and converted to steam. This will increase the pressure inside the piston. The work of the boiler which was converted to heat and transferred to the piston, will be reconverted to work, raising the piston until the pressure is equalized. Pressure varies directly with temperature and inversely with volume. By utilizing the pressure gradient, random kinetic heat energy is rechanneled into motion in the direction of the displacement. Even after removing the piston from the boiler, it will continue to rise until the pressure gradient is neutralized. Now let the atmosphere cool the piston, creating a pressure gradient in the opposite direction. The piston is ready to turn an engine.

The maximum efficiency of the system can be measured by

$$\frac{T_2 - T_1}{T_2},$$

the ratio of the temperature gradient to the first temperature. Obviously the efficiency increases as T_2 increases and T_1 decreases. [Figure 9]

It is acceptable to substitute heat exchanged for the temperature gradient. Entropy, then, is measured by the ratio of the heat exchanged to temperature,

$$\frac{Q}{T_i},$$

in a cycle. What is involved is the net balance of heat exchange to work. The Second Law of Thermodynamics states that the entropy will always be either positive or exactly zero in any complete cycle. The work done by the system will either be less than or equal to the energy introduced into the system. Free energy can no longer be directly equated to the energy introduced into a system. The free energy of a system, or, by this reckoning, its potential capacity to do work, may be less than the energy introduced into it. It will never be greater.

Hence, although entropy theory does not directly impinge on Maxwell's "classical" field theory, its implications are directly contradictory to Riemannian energy potential field theory.

The metaphysical implications of the Second Law are obvious. The world is running "dowwwwn." Dynamic equilibrium contains the inherent entropic tendency to become static.

There is nonetheless an appropriate non-Malthusian generalization to be made on the basis of the Second Law of Thermodynamics. The universe does not have a built-in tendency to heat death, but systems which operate by apparently simple-reproductive cycles do manifest an entropic tendency. They wear out and run down. It is the fact that the universe is not fundamentally cyclical or simple reproductive, but developing, which is critical.

Entropy reflects the tendency of a given system to average out its differences — to depotentiate. If those qualitative small-order differences which signify the potential for a higher order of potentiation, a changed potential function, are not realized, the universe or any subsystem within it can and will at any time suffer an entropic fate. It is incumbent upon us to speedily develop the capability to utilize nuclear fusion power as an efficient source of energy because, otherwise, we will inevitably degrade existing energy sources without the possibility of replenishing them.

The problems inherent in the development of a fusion-based technology illustrate the final points which must be made. At the moment,

Figure 9

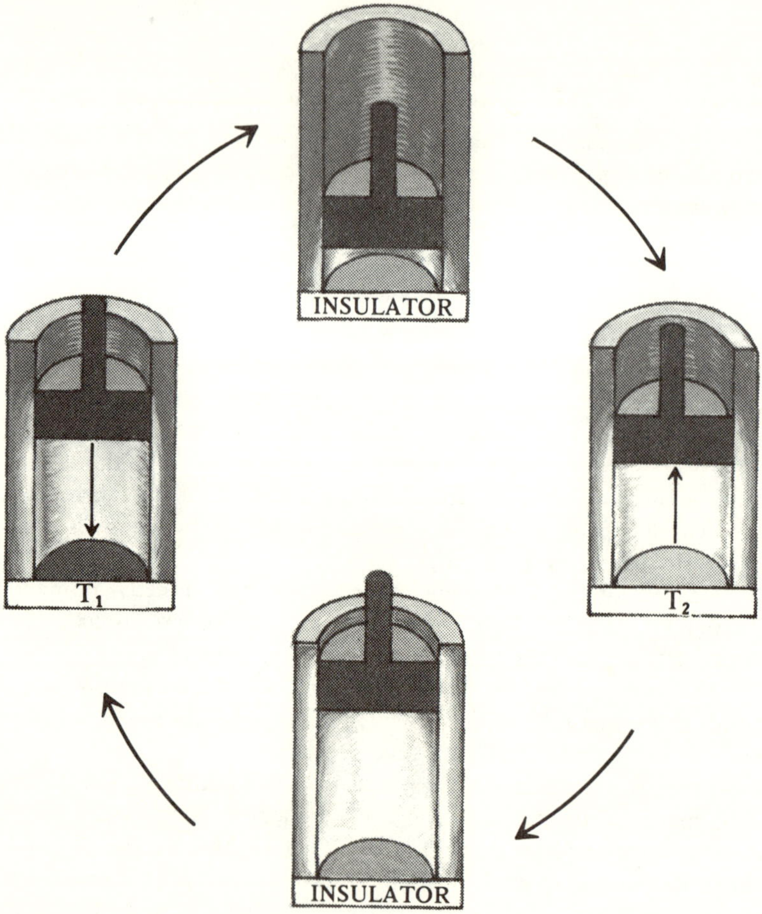

The Carnot cycle which shows an idealized version of the four cycles of a steam engine. Heating and cooling the gas in a cylinder alternately causes the piston to rise and to fall.

---

physicists are working for energy breakeven. By this is simply meant the attempt to achieve a balance between the amount of energy immediately absorbed by hydrogen atoms in order to bring fusion about and the amount of energy released by them after fusion occurs. This is already on the horizon. The serious problem which faces us is the necessity to build the technical and economic infrastructure upon which conversion to fusion can be based.

We have to maximize the rate of energy throughput on the basis of existing resources. We need a crash infusion of free energy into the system, even at the cost of sharply depleting existing coal, oil, and fissionable resources. We face the serious problems of developing materials which can sustain huge temperature gradients ranging from the fraction-of-a-degree temperatures of super cool fluids to the hundreds of millions of degrees of the typical plasmas. We must absorb an enormous quantity of free energy and stretch our existing system to its limits without violating those constraints which would destroy it. We must increase the immediate rate of entropy, an inherent byproduct of such a crash program, in order to nullify that entropic growth by the negentropic emergence of a qualitatively new characteristic universal energy with new, higher-order constraints.

In other words, we must create just those metastable conditions which the French materialist school would consider cataclysmic.

This same higher ordering of energy potentiation occurs spontaneously in high energy plasmas. Under circumstances of magnetic confinement, vortical filamentary structures will arise which circumvent the constraining boundary values of the "Alfven limiting current." The "shock heating" which might be expected to occur in temperatures in the range of 100 million degrees does not occur in the plasma focus current sheath. The current sheath corrugates and the directed energy that would ordinarily be degraded to entropy (thermal energy) in a planar snowplow or shock appears as rotational energy and local magnetic energy of the vortex filaments.

Thus, not only do energy potentiation and social reproduction represent the same universal process, but both can only be determined according to the principle of negentropy.

CHAPTER II

What Is Field Theory — The Particle And The Field

The single most important discovery in field theory before Riemann's work was the development of a concept which would allow the measurement of free-energy density throughput in a system. To do this, it was necessary to conceptualize the energy of a system as a primary, rather than as a derived quality. Initially, force was treated as dependent on the relationship of charged particles; now, that relationship, further generalized as energy, was to subsume the disposition of the particle. In this way, it was possible to approximate the geometry of a field by an energy function.

The notion of universal lawfulness demands at least an approximation to an understanding of energy throughput. Only upon such a basis is it possible to establish appropriate criteria for the investment of resources or the broader moral-political decisions which follow from such criteria. Such political decisions demand exactly the same scientific rigor as is appropriate in the laboratory. The notion of the energy function developed by the French materialists, who were otherwise involved with the French Revolution and, by acquaintance with Benjamin Franklin, with the American, was a necessary tool both politically and in more narrowly "scientific practice." Riemann's critical treatment of the limitation of physiocratic potential theory created the conditions for an actual unified field theory.

Consider the function

$$\frac{Q_1Q_2}{R^2},$$

which expresses the force two charged particles exert on each other. Q_i represents quantity of charge; R represents the distance between them.

Then, Q_i/R expresses the potential which one of these charges, when isolated, possesses of influencing any other charge brought into its neighborhood. Generalize this further and superpose the energy potential of all the paired relationships which you can determine to exist between the particles which determine the field. Now abstract from these charges and consider the field which the charged particles have produced. You have defined that field, by approximation, to be the linear sum of the individual energy relationships existing on a pairwise basis. You know that this cannot be a complete description of the reality, nonetheless, you have made a major step forward. You can now approximately characterize the field as a whole in terms of its energy and begin to penetrate the geometry of the field. This treatment immediately suggests the need for a higher-order geometry which will connect the metastable geometry of the electrostatic field to the "stable" geometry of the particle which produces this field and is determined by it. Immediately, the discovery of even such a linear potential function introduces the geometric conception of potential gradient.

From this point of view, a particle is not propelled toward or away from other particles directly by their action on it, rather, the particle "rolls" down a potential gradient. For the purposes of description, it is useful to consider electrical repulsion as positive, and gravitational and electrical attractive force as negative quantities. Thus, oppositely charged particles which attract each other are moving from positions of higher to positions of lower potential, just as negative 20 is a smaller number than negative five.

Now, lines of equal potential, like contour grading on a hill, define the path of flow, known in fluid dynamics as streamlines. The streamlines will be perpendicular to the lines of equipotential, thereby maximizing the gradient of "descent." An appropriate image is the path and speed of a ball released on a hilltop where it has been constrained to rest. It will travel from a position of high gravitational potential to that of low potential on the steepest path available to it. If we constrain it so that it can only travel on an alleyway, its speed will vary as we vary its path. By adjusting the alleyway from a vertical to a horizontal position we will reduce its speed until, finally, the ball will not move at all. The alleyway will be tangent to a line of equal potential at this point. [Figure 10]

Potential Energy in Field and Particle

We begin our discussion with static fields, although, as we have seen, such a notion ignores the existence of torque and magnetic force equally. The simplest field expresses the relationship between particles on the basis of pairwise interactions through a law of central forces. Thus, the law of

Figure 10

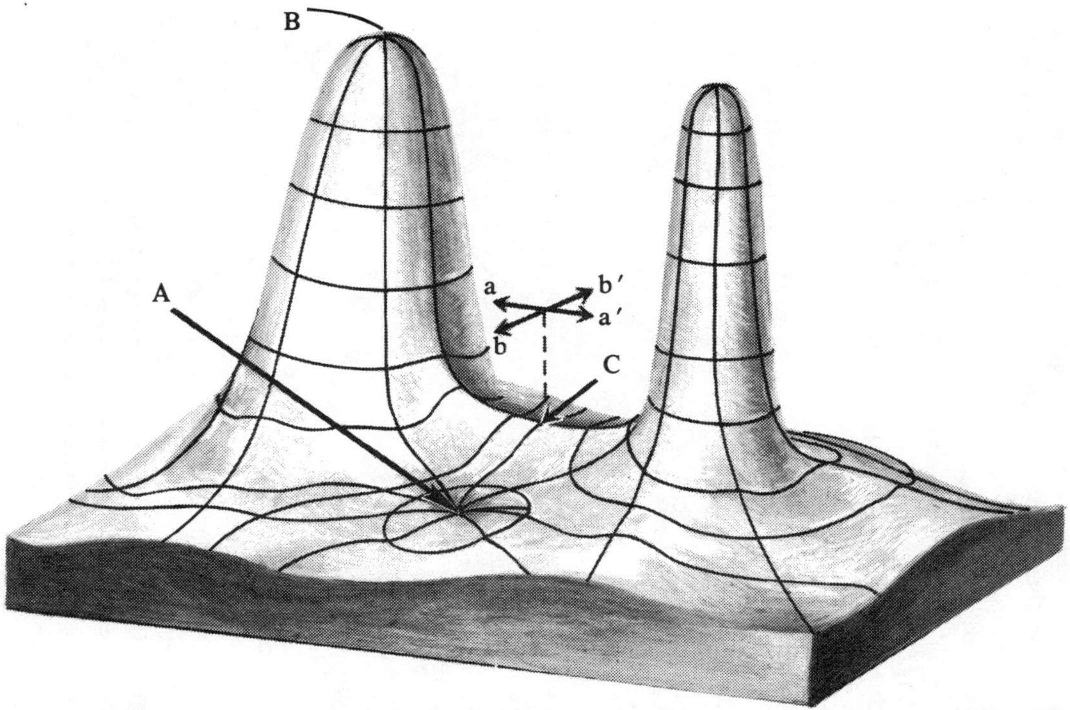

A gravitational potential surface, which may be thought of as a real surface. A particle placed at A, B, or C remains at rest; a plane tangent to any of these points is horizontal. A particle at A, if slightly displaced, returns to A. At B, a slightly displaced particle tends to increase its displacement. If displaced at C in the aa′ direction, the particle will return, but in the bb′ direction, it will not. The three cases at A, B, and C are, respectively, stable and unstable equilibrium and a saddle point.

gravity asserts that two particles are attracted to each other along the line of their connection, directly as the product of their masses and indirectly as the square of their distance. The same law defines the relationship of charged particles. Potential energy may be determined then by the amount of work needed to separate the particles, to "lift" one from the other. The greater the distance a given particle is lifted, the greater its potential to accomplish work by falling.

This quantity may be determined experimentally by measuring the work which would be needed to carry the particle from any given point to a point infinitely distant (in practice this means a point at which the force is no longer appreciable). This can be computed as

$$-G\int_r^\infty \frac{m_1 m_2}{r^2}dr = -G\frac{m_1 m_2}{r}.$$

The potential energy will then be equal to the product of the two masses divided by the distance which separates them:

$$-G\frac{m_1 m_2}{r}.$$

If the body which we are "lifting" has a mass of one, the potential is quite simply expressed as the sum

$$-G\frac{m_1}{r_1} - G\frac{m_2}{r_2} - \ldots$$

or the integral

$$-G\int\frac{dm}{r},$$

if the mass is continuously distributed over a given volume v. As r increases, $-m/r$ approaches zero (but increases since it is a negative number). As r decreases, $-m/r$ becomes smaller as it approaches negative infinity.

The same method is used to determine the field between two charged particles, except that, for the sake of consistency, we define the potential energy of like charges which repel each other as positive.

We can now establish the potential function V as

$$-G\int\frac{dm}{r}$$

or

$$\int\frac{dQ}{r}$$

by superposing pairwise relations between the particles of the collectivity and a test particle of unit mass or unit charge. By simply reversing our original line of reasoning, we can now calculate the force as the gradient of the function V. The gradient can now be expressed by partial derivatives as

$$\mathbf{i}\frac{\partial V}{\partial x} + \mathbf{j}\frac{\partial V}{\partial y} + \mathbf{k}\frac{\partial V}{\partial z}$$

where **i**, **j**, **k** are unit vectors in the x, y, and z directions.

In 1781, Lagrange discovered this relationship. Laplace quickly showed that the second partial derivatives of V satisfied the equation

$$\frac{\partial^2 V}{\partial x^2} + \frac{\partial^2 V}{\partial y^2} + \frac{\partial^2 V}{\partial z^2} = 0,$$

subsequently known as the Laplacian. (The second derivative repeats the operation of differentiation twice. V here denotes the potential energy so that

$$\frac{\partial^2 V}{\partial x^2}$$

is the rate of change of the rate of change of potential energy in the x spatial direction.)

In 1813, Poisson made the crucial correction that when the point is within the substance of the attracting body, rather than in the space surrounding it, the Laplacian equation must be replaced by

$$\frac{\partial^2 V}{\partial x^2} + \frac{\partial^2 V}{\partial y^2} + \frac{\partial^2 V}{\partial z^2} = 4\pi\varrho;$$

where ϱ denotes the density of mass or charge.

Gauss and George Green, in the following decade, established the conditions under which this function V was unique. In any given case, establishment of the uniqueness of the potential function is equivalent to the proof that there is conservation of energy. If more than one function describes the potential energy of the system, then, at a particular point, an object could be subject to two or more different motions, giving rise to a multiplicity of kinetic energies for the object and a lack of definiteness in the total energy content of the system. If a system is conservative, then any complete cycle must restore conditions to their initial value. If a pendulum does not achieve the same height at the end of a swing as at the beginning, then the system has suffered an increase of entropy and lost free energy.

Contrast the pendulum to a paddle wheel placed at a bend in a stream in such a way that an axle attaches its center to the stream bed. The stream will continuously work on the wheel causing it to rotate. A 360 degree turn of the wheel, a complete cycle, will not return an equivalent amount of energy to the stream. (The paddles can be flexible so as to minimize the return of work to the stream.) This is a case of rotation, or curl as it is sometimes known. In this case, energy will not be conserved and a unique potential function does not exist. Every complete cycle will result in an addition of a fixed amount of work to the wheel by the amount of the product of 2π and a constant. This work may be stored by the wheel as increased velocity. Some will be dissipated by frictional heat. Free energy may be introduced into the system.

The notion of the potential function was merely preliminary to the necessary higher-order concept of energy density. The divergence of energy density is that concept which allows us to connect, at least by approximation, the particle and the field. Since the particle density can be expressed in relation to the potential energy function, the particle (or energy resource) is implicitly defined in terms of its energy potentiation. Although twentieth century considerations of the potential energy stored in the nucleus of an atom demonstrate that the apparently explicit definition offered by classical electromagnetism is incorrect, the rudimentary definition of the energy density by Poisson was a major historic achievement.

The uniqueness of the potential function is an implied consequence of Laplace's original formula. It was Poisson's correction which provided the basis for a precise mathematical description of energy throughput. The easiest way to understand the concept of energy throughput is by transforming the problem to an equivalent one in fluid dynamics. It is essential to bear in mind that the notions of free energy throughput, which we are now describing, have the previously identified physiocratic flaw. They do not allow for the qualitative emergence of new forms of energy — a "surplus" over and above the energy previously contained within the system.

For purposes of this discussion, we can compare the free energy to a fluid which moves through a tube. The velocity of its flow is equivalent to the attractive force of gravity or the repulsive force of two like-charged electric particles.

The Laplacian,

$$\frac{\partial^2 V}{\partial x^2} + \frac{\partial^2 V}{\partial y^2} + \frac{\partial^2 V}{\partial z^2},$$

gives us a measure of divergence. It tells us whether the fluid is compressible, that is, whether its density can be increased. If the Laplacian is zero, then, when we decrease the volume of the tube, the velocity of flow will increase, and conversely, an increase in the volume will be compensated by a slowing of the flow. The divergence of the fluid is zero. There is no change in density of the fluid as it flows through the tube. Energy throughput in the system is constant.

Poisson's formula tells us that the Laplacian is not equal to zero at a point within an attracting body, but instead has the value $\pm 4\pi\varrho$, where ϱ represents the density of charge. Here, or course, any meaningful analogy to fluid dynamics breaks down. But, the way in which it breaks down is relevant. The charged particle is viewed as either an unlimited source of fluid or a repository sink, since the attractive or repulsive force between two fixed charged particles does not abate. The fluid will pour into the given region at the rate of $4\pi\varrho$ cubic units of measurement per second. (ϱ, m, or Q are used here interchangeably, referring to density in general, mass or charge in particular.) Therefore, the quantity of charge is equal to the change in velocity of the imaginary fluid at the boundary of the charge.

$$\frac{\partial^2 V}{\partial x^2} = \frac{\partial E}{\partial x},$$

where V denotes energy potential and E denotes electric force.

The derivation of Poisson's correction of Laplace's formula for the divergence of the field at a charged particle exposes the inherently paradoxical nature of classical particle-field approximation to a potential energy theory. The problem arises as soon as we remember that we are dealing with a collectivity. In reality, we never deal directly with charged particles, with single electrons, but with material bodies on which charge has collected. Then, these bodies must have internal interactions. How, then, do we describe their internal field? Without such an internal field, how can we account for the ability of charged particles to effect each other? Yet, at the boundary of a charged particle, R, the distance between the particles, is equal to zero and the potential field would appear to be infinite in strength, having R in the denominator. Mathematically, such a point is know as a singularity. Practically, it raises the question: Why does a body, which is charged, not explode under the repulsive force of the forces between the charged particles collected on it? Clearly, the notion of material binding forces which act as a cohesive force to hold matter together begs the question. The very separation of the electromagnetic field and the gravitational field is a clue that something is amiss. We lack a unified field theory which can subsume the real existence of the particle as a metastable, highly ordered dynamic complex of the field.

Traditionally, one of the most serious problems in mathematical physics has been posed by determining how such boundaries can be crossed so as to approximate measurement. This is a useful, in fact, highly necessary task, particularly as such points of singularity are rigorously identified. It is the consideration of just these singularities at the borderline of the infinite which defines the necessity of new higher-order manifolds which subsume the areas of separation and the boundary lines which separate them in a new functional relationship.

The boundary introduces the necessary nonlinearity which explodes the apparently rule-ordered behavior which exists dichotomously in descriptions of the field outside the particle and the field inside the particle containing two different functions:

$$\frac{\partial^2 V}{\partial x^2} + \frac{\partial^2 V}{\partial y^2} + \frac{\partial^2 V}{\partial z^2} = 0$$

and

$$\frac{\partial^2 V}{\partial x^2} + \frac{\partial^2 V}{\partial y^2} + \frac{\partial^2 V}{\partial z^2} = 4\pi\varrho .$$

The derivation of the field of a charged particle is not difficult pragmatically. The rate of change of the force, the divergence, will differ internally and externally, but the force is continuous at the boundary of the particle. In practice, the offending zero denominator is eliminated by enclosing a charged particle by an infinitesimally small sphere. The area of a sphere is $4\pi R^2$. Every part of the area of the sphere will be subject to the force emitted by the particle. Multiply the area by the force Q/R^2 and the result is $4\pi Q$. In this way, the charge is measured by the force it exerts on the field as a whole, rather than in relation to another charge per se. It is worthwhile to note that the second derivative is not only represented by different functions internally and externally, but is discontinuous at the border, emphasizing the second-order nonlinear effects which can only be ignored by first approximation.

We can, if we wish, now think of this force as flowing out of the small sphere at a rate of $4\pi Q$ units per second. This relationship of the flux of force through a surface and the density of charge contained within it is known as Gauss's or Green's flux law (having been discovered independently by each man). [Figure 11]

An Appropriate Epistemology— Negentropy

Now we are in a position to consider the question of the measurement of the negentropic potential of energy density throughput. Do we have a way to measure the potential of a given system to increase its energy density?

The following remarks from a speech given by Soviet physicist, P.L. Kapitsa, to the Soviet Academy of Science in 1975 broaden the concept of

Figure 11

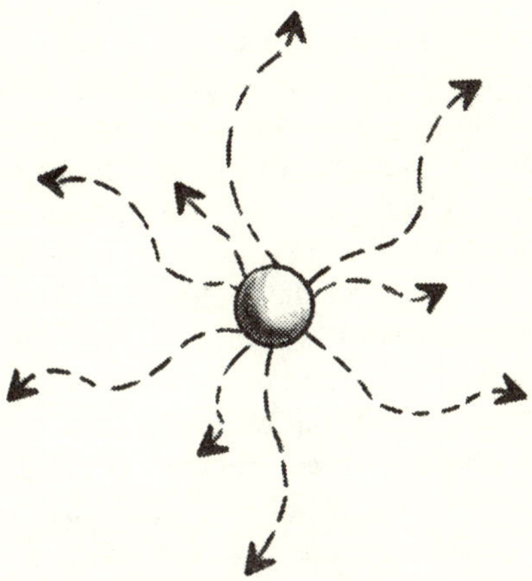

The flux of force out of a charge. It should be clearly understood that nothing is flowing out of the charge. The flux notion does, however, help to clarify the relation between the charged particle and its field.

energy density along these lines. We quote from a translation appearing in the June 1976 *Fusion Energy Foundation Newsletter:*

> All energetic processes of interest to us boil down to the transformation of one type of energy into another, and this takes place according to the law of conservation of energy. The most commonly used forms of energy are electrical, thermal, chemical, mechanical, and now so-called nuclear. The transformation of energy can usually be viewed as taking place in a certain volume, through whose surface one form of energy enters and transformed energy comes out.
>
> The density of the yielded energy is limited by the physical properties of the medium through which it flows. In a material medium, the power of energy flow is limited by the following expression: $\mathbf{U} < \gamma F$, where γ is the velocity of diffusion usually equal to the speed of sound, F can be either mechanical or thermal energy, and **U** is a vector. In stationary processes, divergence **U** (the variation of energy flux from place to place in the medium) determines the magnitude of energy transformation into another form.

Kapitsa succintly fuses the definition which is adopted, in practice, in making decisions about industrial planning with that in scientific use. The notion of net energy throughput is combined with that of energy density to give us the measure of net free energy density. While Kapitsa does not explicitly contradict the traditional use of the law of conservation of energy in this piece, the implications are obvious.

Immediately, nuclear physicists are at the point of net energy breakeven in fusion reactions. This means that as much energy can be released in a controlled reaction as is immediately introduced into the system to trigger the reaction. This measure is simply a benchmark which demonstrates the feasibility of controlled thermonuclear reactions and justifies the commitment of the resources it will take to make the necessary industrial transformations implied in its use as a global energy source.

For purposes of planning such a transformation, a broader notion of net energy transfer is needed, which includes an actual accounting for fixed capital depreciation and variable capital throughput (men and materials involved in online production) and various associated social costs.

It is clear from examples already at hand, for instance in the direct production of electricity by the reduction of coal to a plasma state (magnetohydrodynamics or MHD), that such questions of cost are not problematic. We can expect free energy density to be well above breakeven; however, the question of industrial planning immediately emphasizes the inadequacy of a notion of net energy density which does not include the whole cycle of production. From that point of view, the cost of introducing fusion power is balanced against the profits from increased surpluses in every other sphere of industry and agriculture which will arise as the costs of production and distribution in general are lowered. Whole new ranges of product become available. Free energy density properly is a measure of the throughput of energy from one cycle to the next which measures the net energy transfer in any given cycle by the potentiality which it establishes for a higher order of throughput in the next cycle. The negentropic process is defined by the tendency for the rate of change of net energy density to itself increase.

The Photoelectric Effect

Maxwell failed to see the epistemological implications of Poisson's discovery. Instead, he dismissed the notion of singularities as mere inconveniences which could be ignored without peril. A singularity is a region in which a function fails to maintain continuity, either by jumping between values without taking on the intervening values or by becoming infinite.

Generations of students of quantum physics now suffer the pangs which follow such cavalier ingestion of sour apples. The first signs of indigestion occurred when the anomolous photoelectric effect was observed.

To go ahead of ourselves, Maxwell established the position of electromagnetic waves in the radiation spectrum by recourse to a field theory which treated the field as a literal material medium, an aether. The particles of such a field were aligned in reaction to the advent of charged particles; waves were propelled along them as these particles varied their position or oscillated. The analogy is clearly to surface water waves. In Maxwell's view, it was not necessary to develop a theory which accounted for a higher-order geometry of the field within and without the particle. He rejected the notion of energy potential as merely a useful mathematical fiction. Just as it is unnecessary to know the character of a rock to describe the waves which follow upon impact when it is thrown into the water, so Maxwell treated radiation waves. The energy of radiation waves was assumed to directly relate to the impact of the particle oscillating on the aether, just as the water wave depends upon the amplitude of the displacement caused by the impact of a rock thrown on its surface.

Although water is made up of molecules and is not, therefore, fundamentally continuous in Maxwell's terms, waves on the surface of a pond spread out through a continuous motion. They do not transmit the rock's impact by turning themselves into a succession of discrete water-pebbles. Another way of expressing this is to assert that the divergence of a radiation field is zero except at the location of a charged particle.

The notion that the divergence is zero in an electric field in which there is no energy source present is a critical feature of Maxwell's theory and the basis on which he deduces the existence of electromagnetic waves. (As we shall later show, the form in which he presents his theory is not necessary to his conclusions.) Empirical proof that the electromagnetic field can not have zero divergence at any point came only in the twentieth century with confirmation of Einstein's photon hypothesis.

The electromagnetic field, in effect, becomes pebbled at the same time that it propagates waves. That is, it has a highly complex internal geometry as considerations of negentropy would suggest. Only the subterfuge of a material aether, which would propagate continuous waves, made Maxwell's approach even plausible.

To discuss Einstein's hypothesis, it is necessary to presume the reader's familiarity with the notion of light as a wave phenomenon and with the range of the spectrum which places ultraviolet light at the upper end with a frequency approximating to 10^{16} cycles per second. As early as 1887, Hertz discovered the photoelectric effect by which a metal surface bombarded by light will emit electrons. In a cathode ray tube, electrons are emitted from a metallic negative electrode (the cathode) as a result of the bombardment of the electrode by positive ions of the gas contained in the tube. Hertz's apparatus is shown in Figure 12. The glass tube contains a polished metal electrode, called a photocathode, in the form of a perforated metal plate. The two electrodes are maintained at a potential difference of a few volts (normally with the second electrode positive with respect to the photocathode). When ultraviolet light passes through the perforated electrode and is incident upon the inner surface of the photocathode, a current is observed to flow through the tube. This phenomenon is called the photoelectric effect. It persists even when the tube is evacuated to a very low pressure. This implies that the current is not carried by the motion of gaseous ions (a convection current). Experiments in which a magnetic field was applied to the region between the photocathode and the second electrode indicated that the current consisted of the flow of negatively charged particles.

The Lorentz force law asserts that an electron traveling through a magnetic field will be deflected in a direction perpendicular to its original velocity and the direction of the field. Consider that the magnetic field is established in a vertical direction and that the particle flows from east to west across the sheet of paper on which Figure 12 is drawn. If the particle is positive, it will be deflected in a southerly direction. If it is negative, it will experience a force to the north. The hypothesis that these negatively charged particles were electrons was confirmed by Lenard in 1900 when he established that they had the same ratio of charge to mass.

The experiments of Lenard cleared up the question of the identity of the photoelectric particles, but they also demonstrated some properties of the photoelectric effect which, as we shall see, were very difficult to understand in terms of the theories of classical physics. Lenard measured the current reaching the second electrode as a function of the potential between the electrode and the photocathode, keeping all the other parameters fixed. His data are represented in Figure 13. An interesting feature of the data is that some current reaches the second electrode even when the second electrode is given a negative charge. Remember that like charges repel each other, therefore, you might expect that the negatively charged electrons would be repelled away from the negatively charged second electrode.

The conclusion to be drawn is that the electrons are ejected from the photocathode with sufficient kinetic energy to overcome the repulsive force. There is a well-defined point beyond which the negative charge of the second electrode cannot be increased without the current ceasing to flow. This is the point at which the kinetic and potential energy balance each other. This is denoted by $-V_{max}$. The decrease in current up to that point indicates that there is variation in the kinetic energy of electrons and

Figure 12

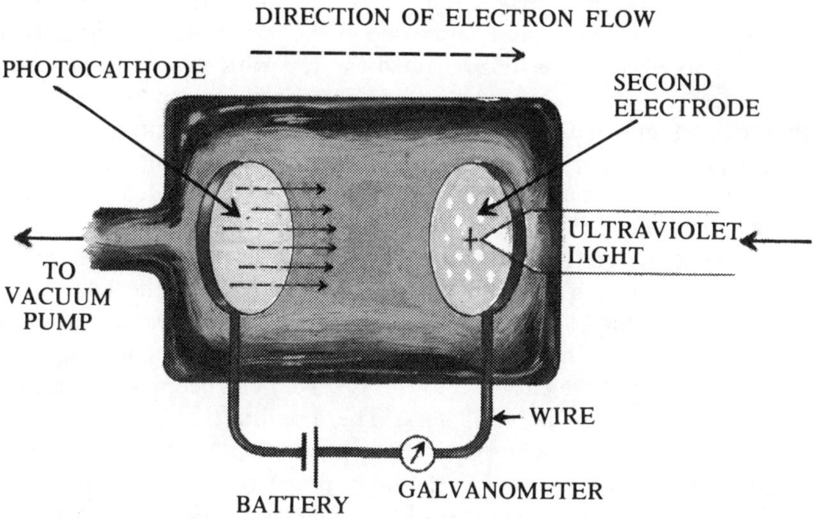

Light passes through the perforated second electrode to the photocathode. Electrons are discharged from the photocathode and flow in the direction of the second electrode, since the second electrode is maintained at a positive potential relative to the photocathode.

Figure 13

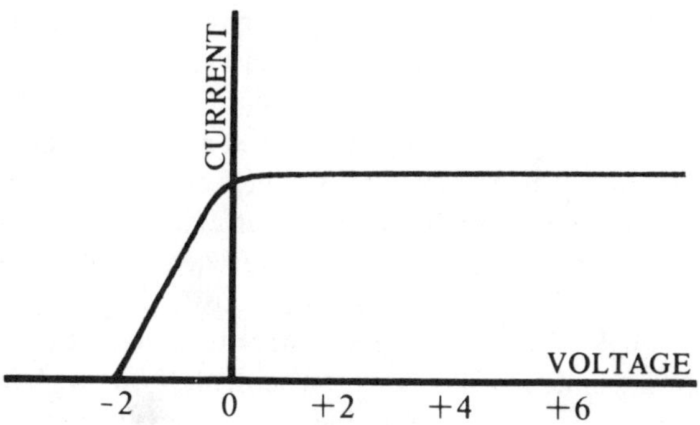

Potential of the second electrode with respect to the cathode.

a likely hypothesis is that those electrons nearest to the surface have maximum energy since they meet with minimal resistance.

Thus far in his experiment, Lenard had varied the potential difference and discovered an expected direct correlation in the variation of the current. Now he tested the relationship of the current to the intensity of the ultraviolet light. He first repeated the test with a positive potential difference between the second electrode and the photocathode and found that the current was directly proportional to the light intensity. This would have led him to expect that the same relationship would hold with a negative potential difference, so that, with a less intense incidence of ultraviolet light, the cut-off point of $-V_{max}$ would occur sooner; that is, that less negative potential difference could be sustained. Surprisingly, this was not the case: $-V_{max}$ proved to be independent of the intensity of the light.

Maxwell's theory would have predicted that the interaction between the electromagnetic light waves and the electric charge of the electrons would have caused the electrons to vibrate. They can be imagined as little pendulums or harmonic oscillators. The amplitude of their vibration, according to Maxwell's theory, would be proportional to the amplitude of the electric field, which, in turn, should be proportional to the square root of the light intensity. The energy of the photoelectrons should have been dependent upon the intensity of the light, yet experiment showed that it was not.

Another consideration makes the problem even more serious. A comparatively long time should be required for an electron to absorb enough energy to overcome the binding force of the photocathode and the resistance of the dielectric. This also is not born out by experiment. The predicted time would be in the order of 100 seconds. Observed time was not higher than 10^{-8} seconds.

Einstein's photon hypothesis explains these anomolies by assuming that the energy emitted by the ultraviolet light is not continuously distributed over the field, but groups itself in discrete light quanta which he named photons. The energy of a photon would be proportional to the frequency of the light and a constant, h. He also assumed that in the photoelectric process an exchange occurs in which one of the quanta is completely absored by an electron in the photocathode. The energy of an individual photoelectron is proportional to the frequency of the light, not its intensity. The rate of emission of photoelectrons is proportional to the intensity of the light. This is in agreement with Lenard's observations. It explains the independence of $-V_{max}$. The theory also explains why there is not a time lapse of over a minute before a photoelectron is released. A uniform electric field, one with zero divergence, would have a continuous potential function. Energy would be spread uniformly over the entire wave front. However, as experiment shows, the energy distribution is not continuous; the field is not comparable to an incompressible fluid evenly spread, even in those regions in which sources are not present. When a single photon strikes the photocathode, it will transfer a sharply localized, concentrated quantum of energy. The energy of the expelled photoelectron is solely a function of the frequency of the incident radiation; only the number of photoelectrons expelled depends upon the intensity.

The fact that the oscillating electromagnetic field develops a structure which concentrates energy discontinuously as a function of its cyclical variability can only be problematic to someone who imagines that an unlimited source or sink is an appropriate conception of a charged particle. A field can only properly be defined as a collectivity which is inclusive of the particles which produce it by their interaction, and which develops new degrees of freedom within higher-order manifolds and increased density of energy throughput as a periodic function of its collective interaction.

The quantization of energy $E=h\nu$, where ν is the frequency of oscillation, demonstrates that free energy density is not simply a function of a fixed charged-particle resource, but is also a dynamic function of the system. Even with reference to the apparently continuous electromagnetic radiation field, it becomes possible to develop a standard of measurement. The measurement of negentropy is embedded within the topology of the universe. Thus, the possibility of an empirical solution to the problem raised by Riemann in the ***Hypotheses Which Lie at the Foundations of Geometry*** opens up.

> Determinate parts of a manifold, distinguished by a mark or by a boundary, are called quanta. Their comparison as to quantity comes in discrete magnitudes by counting, in continuous magnitude by measurement. Measuring consists in superposition of the magnitudes to be compared; for measurement there is requisite some means of carrying forward one magnitude as a measure for the other. In default of this, one can compare two magnitudes only when the one is a part of the other, and even then one can only decide upon the question of more and less, not upon the question of how many...

Measurement of negentropy depends upon the comparison of successive cycles during which the universe reproduces itself. In each succeeding period, its topology becomes increasingly more complex, of a more highly ordered energy potential gradient. The development of a photon structure in the radiation field, just as the development of vortex filaments in energy dense plasmas, indicates the inherent universal tendency to develop higher-order energy structures in direct contradiction to the Second Law of Thermodynamics.

The notion of progress implies that distance and time are relative, but not indifferent concepts. The real measure of the space-time between the United States and Europe is no longer the two months necessary to cross the Atlantic by sail, but incorporates the time of air travel and telecommunications as included notions. A comparison of the "shrinking" of distance over the past 200 years makes the point.

The comparison of energy density throughput provides the absolute measure not only for space-time, but for material objects. This is recognized in honest industrial accounting practices which recognize that fixed capital depreciates at a quicker rate than its useful life. Fixed capital is properly accounted for from the point of view of social reproductive potential, by its cost of replacement rather than the original cost of investment. With advancing technology, this is constantly being cheapened. It is exactly this "cheapening" of fixed capital which properly places an enhanced value on invention.

In any given cycle of production, the absolute density of free energy throughput must be increased by the standard of preceding periods. This is implied by the expansion of production, higher living standards, population growth, and increase of leisure which are the criteria for progress. However, it is by reference to the future that the admissible rate of such expansion of throughput is determined. It is in the potentiation for future expansion of potential that present valuation is properly adduced.

CHAPTER III

What Is Field Theory —Action-At-A-Distance

Maxwell and Hertz perpetrated the fraud that Michael Faraday discovered field theory because he treated action in the small and that the Maxwell-Faraday theory represents the only genuine such theory. This blatant disregard of Riemann's writings on the subject, available to both writers and published in the second section of this book, can only be viewed as outright dishonesty or hysterical blocking. Riemann developed the simple notion of the retarded potential, and elaborated and modified Weber and Gauss's theoretical work in electromagnetism in terms of this notion.

Simply put, retarded potential describes the obvious fact that a potential field is not immediately established between two charged particles. It develops with the speed of light. This is extremely, but not infinitely fast.

The word retarded merely implies that at any given point in the field the potential can only be determined according to the position and velocity of the collectivity of particles in the immediately preceding period. In other words, action-at-a-distance does not occur instantaneously, but in time. On the basis of this notion of a potential field developing in time, Riemann anticipated Maxwell's actual contributions without the accompanying epistemological blunders.

While it is true that he too looked at the progressive development of a field, Riemann never counterposed his work to that of Gauss and Weber as did Faraday. Faraday conceived of himself as carrying on a heated, if one-sided polemic with continental physicists who were relatively uninterested in his epistemology. The notion of action-at-a-distance was regarded by continental physicists of Faraday and Maxwell's time as merely a convenient conceptual fiction for an infinitely fast velocity of propagation. While it was broadly suspected that the electric and magnetic fields were propagated with the speed of light, the empirical and mathematical foundations for superseding that convenient approximation or fiction were not available to them. Even Newton, in his letter to Bentley, disclaimed any but heuristic value to his theory of gravitational action-at-a-distance.

Faraday's epistemology, whereby he claimed to establish the reality of the inbetweeness of the field through which electromagnetic action is propagated, is flawed by his deliberate exclusion of the particle from the field and by the same problems which afflicted Zeno's tortoise. It will be useful to treat the subject of the electrical field as it developed historically in order to subject Faraday's view to criticism.

The Electrostatic Field: Faraday and Franklin

For the same reasons that determine the ability of elements to interact chemically, materials in general manifest an electric potential difference when they are placed in contact. Electricity produced by rubbing is an effect produced by potential differences between insulating materials and can result in rather large residual charges.

The lack of conductivity of these insulators means that each small section of one that comes in contact with the other will acquire its own local charge and potential. In order to have any considerable amount of charge transferred, there must be a large area of contact, which normally requires rubbing the two substances together. Amber (in Greek, *elektron*) was the earliest substance known as an electric. Amber, when rubbed against silk or cat's fur, will become negatively charged. The fur will be positive. Glass rubbed by the same fur will become positive, leaving the fur negative. Resins will become negative when rubbed against glass or amber. (Benjamin Franklin gave the name "negative" and "positive" to what had previously been called resinous and vitreous electricity on the mistaken assumption that it was the positive rather than the negative charge which contained mobile free electricity.)

The pictorial representation of field lines traveling from the positive to the negative charge arises from the analogy of an electric field to a fluid. The assumption which underlies this analogy is that a subtle, imponderable, electric fluid flows out from a positively charged substance to a negative one. The positive charge acts like a source; the negative charge a sink. This assumption can lead to the kind of disastrous banality which afflicts many textbooks. The field is pictured as having a direction, flowing from positive to negative. Thus, the field between an electron and a positive charge will be represented by arrows pointing toward the electron. But, as we shall show, the electric field does not have the property of vector directionality. It is a geometric transformation of space as a whole, a shrinking or stretching which depends upon whether attraction or repulsion is occurring. There is not a force emanating from one charged particle or polarized substance to another, but a relationship inclusive of both.

The following demonstrations should banish any false notion of space as a composite of linear force vectors. Take a ball of light-weight substance which is coated with a conducting material. Suspend it from a thread and place a rod which has been given a negative charge nearby. The ball will be attracted toward the rod. The field set up between the ball and rod will have induced a positive charge on the near side of the ball, that is, free electrons, which are more mobile than positively charged nuclei, will have migrated to the far side. The positive near side of the ball will exert a greater attractive force toward the rod than the negative far side (the distance R will be smaller). The rod will also be attracted to the ball but, in these circumstances, this is not determining.

Reverse the process. Remove the rod and ground it so that its charge is neutralized. Now give it a positive charge and repeat the experiment. This time, the ball will also be attracted to the rod. The near side of the ball will have received an induced negative charge and the far side will be positive. The direction of the force was in the first example from positive to negative, in the second, the direction is reversed.

Touch the rod to the ball and it will fly away. Some of the electrons collected on the ball will have flowed to the rod, which, being positive, had a deficit of electrons. The attractive force between the rod and ball will no longer exist. Remember, unlike charges attract, like charges repel. Now take the rod away from the ball. Recharge it positively and again bring it near the ball. This time the ball will be repelled away from the rod. By touching the ball, thus allowing electrons to flow from it to the rod, we positively charge the ball. Figure 14 graphically represents the experiments. It was such a line of experiment that led Franklin to his theory of conservation of charge.

There were several competing descriptions of electricity in the eighteenth century. Theophile Deaguilier prepared the following description for the French Academy in 1733:

> Around an electrified body there is formed a vortex of exceedingly fine matter in a state of agitation which urges towards the body such light substances as lie within its sphere of activity. The existence of this vortex is more than a mere conjecture; for when an electrified body is brought close to the face it causes a sensation like that of encountering a cobweb.

The Abbé Jean-Antoine Nollet proposed a theory in a series of papers for the French Academy from 1745 onwards, which was widely accepted. Nollet attributed electric phenomena to the movement in opposite direc-

Figure 14

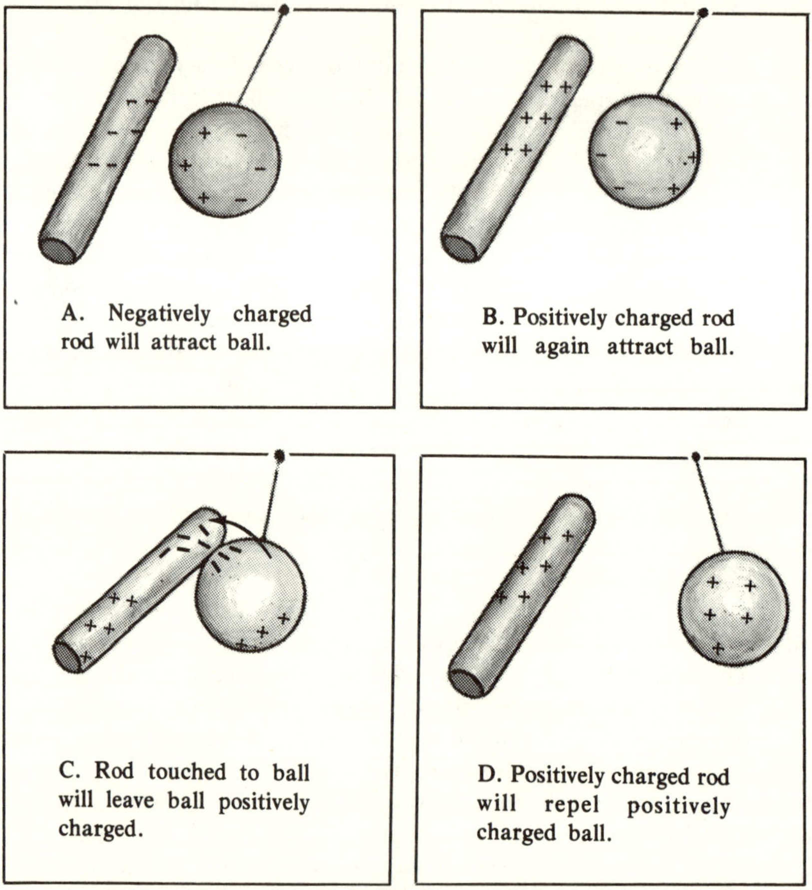

A. Negatively charged rod will attract ball.

B. Positively charged rod will again attract ball.

C. Rod touched to ball will leave ball positively charged.

D. Positively charged rod will repel positively charged ball.

tions of two currents of a fluid, "very subtle and inflammable," which he supposed to be present in all bodies under all circumstances. When an electric is excited by friction, part of this fluid escapes from its pores, forming an effluent stream; and this loss is repaired by an affluent stream of the same fluid entering the body from outside. Light bodies in the vicinity, being caught in one or the other of these streams, are attracted to or repelled from the excited electric. These theories reason by analogy to heat, which at the time was conceived as an imponderable substance.

Franklin was the first scientist to focus his attention on the charge. In so doing, he discovered the principle of conservation of charge. He particularly observed the phenomenon of sparking which occurs when two unlike charges are placed in sufficiently close proximity, without quite touching, so that electrons "leap the gap." He reasoned as follows: If one person A, standing on wax so that electricity cannot pass from him to the ground, rubs a glass tube, and if another person B, likewise standing on wax, passes his knuckle near the glass so as to receive its electricity, then both A and B will be capable of giving a spark to a third person C standing on the floor; that is, they will be electrified. If, however, A and B touch each other during or after the rubbing of the glass tube, they will not be electrified.

It followed that electricity is an element present in a certain proportion in all matter in its normal condition so that, before the rubbing, each person A, B, and C has an equal share. The effect of the rubbing is to transfer some of A's electricity to the glass, whence it is transferred to B. Thus, A has a deficiency and B a superfluity of electricity; and if either of them approaches C, who has the normal amount, the distribution will be equalized by a spark. If, however, A and B are in contact, electricity flows between them so as to reestablish the original equality, and neither is then more electrified with respect to C. Thus, electricity is not created by rubbing the glass, but only transferred to the glass from person A, so that A loses exactly as much as the glass gains; the total quantity of electricity in any insulated system is invariable.

Franklin did not reject the notion of an electric effluvia. He pictured his electricity as an elastic fluid, which repelled its own particles. He described the particles as "extremely subtile, since (electricity — C.W.) can permeate common matter, even the densest metals, with such ease and freedom as not to receive any perceptible resistance." In order to account for the attraction between oppositely charged bodies, in one of which there is an excess of electricity as compared with ordinary matter and in the other an excess of ordinary matter as compared with electricity, Franklin assumed that "though the particles of electrical matter do repel each other, they are strongly attracted by all other matter," so that "common matter is a kind of sponge of the electrical fluid."

He was at a loss to explain the insulating qualities of a dielectric, such as glass, particularly in such phenomena as the Leyden jar which allowed a high degree of charge to collect on two condenser plates. Neither current nor a spark flows through a dielectric, yet attractive and repulsive forces are still exerted. He could only explain this as some form of action-at-a-distance, although he did not envisage his electrical effluvia as necessarily bound within the confines of matter. Thus he writes:

> The form of the electrical atmosphere is that of the body it surrounds. This shape may be rendered visible in a still air, by raising a smoke from dry rosin dropped into a hot teaspoon under the electrified body, which will be attracted, and spread itself

> equally on all sides, covering and concealing the body. And this form it takes, because it is attracted by all parts of the surface of the body though it cannot enter the substance already replete. Without this attraction, it would not remain round the body, but dissipate in the air.

It was Faraday's accomplishment to explain how a dielectric transports electrical effects by itself becoming polarized. Normally, bulk matter exists in an electrically neutral state, but under certain circumstances, when an electric field is applied, a separation of positive and negative charge occurs within the material, leading to the polarized state. It is usual today to brush aside the critically important achievements of potential field theory in order to over emphasize and distort Faraday's achievement. Hertz is typical in this respect. We quote from his *Electric Waves* at some length because his description of essentially separate approaches to the electric field is useful in identifying the confusion of viewpoint which still afflicts textbooks today.

Hertz states the case as follows:

> From the first standpoint, we regard the attraction of two bodies as a kind of spiritual affinity between them. The force which each of the two exerts is bound up with the presence of the other body. In order that force should be present at all, there must be at least two bodies present. In some way, a magnet only obtains its force when another magnet is brought into its neighbourhood. This conception is the pure conception of action-at-a-distance, the conception of Coulomb's law. (The force between two charged particles is Q_1Q_2/R^2, the product of the amount of their charge divided by the square of the distance between them. — C.W.) In the theory of electricity it has almost been abandoned, but it is still used in the theory of gravitation. Mathematical astronomy speaks of the attraction between the sun and a planet, but with attraction in empty space it has no concern.
>
> From the second standpoint we still regard the attraction of the bodies as a kind of spiritual influence of each upon the other. But although we admit that we can only notice this action when we have at least two bodies, we further assume that each of the acting bodies continually strives to excite at all surrounding points attractions of definite magnitude and direction, even if no other similar bodies happen to be in the neighbourhood. With these strivings, varying always from point to point, we will (according to this conception) excite the surrounding space. At the same time we do not assume that there is any change at the place where the action is

exerted; the acting body is still both the seat and the source of the force. This is about the standpoint of potential theory. It obviously is also the standpoint of certain chapters in Maxwell's work, although it is not the standpoint of Maxwell's theory. In order to compare these conceptions more easily with one another, we represent from this standpoint [see Figure 15A] two oppositely electrified condenser-plates. The diagrammatic representation will be easily understood; upon the plates are seen the positive and negative electricities (as if they were material); the force between the plates is indicated by arrows. From this standpoint it is immaterial whether the space between the plates is full or empty. If we admit the existence of the light-aether, but suppose that it is removed from a part B of the space, the force will still remain unaltered in this space.

The third standpoint retains the conceptions of the second, but adds to them a further complication. It assumes that the action of two bodies is not determined solely by forces acting directly at a distance. It rather assumes that the forces induce changes in the space (supposed to be nowhere empty) and that these again give rise to new distance-forces. The attractions between the separate bodies depend, then, partly upon their direct action, and partly upon the influence of the changes in the medium. The change in the medium itself is regarded as an electric or magnetic polarization force. This view has been developed by Poisson with respect to statical phenomena in magnetism and has been transferred by Mossotti to electrical phenomena. In its most general development, and in its extension over the whole domain of electromagnetism, it is represented by Helmholtz's theory.

Figure 15B illustrates this standpoint for the case in which the medium plays only a small part in the resultant action. Upon the plates are seen the free electricities, and in the parts of the dielectric the electrical fluids which are separated, but which cannot be divorced from each other. Let us suppose that the space between the plates contains only light-aether and let a space, such as B, be hollowed out of this; the forces will then remain in this space, but the polarization will disappear.

One limiting case of this mode of conception is of especial importance. As closer examination shows, we can split up the resultant action (which alone can be observed) of material bodies upon one another into an influence due to direct action-at-a-distance, and an influence due to the intervening medium. We can increase that part of the total body which has its seat in the electrified bodies at the expense of that part which we seek in the medium, and conversely. Now in the limiting case we seek the whole of the energy in the medium. Since no energy corresponds to the electricities which exist upon the conductors, the distance-forces must become infinitely small. But for this it is a necessary condition that no free electricity should be present. The electricity must therefore behave itself like an incompressible fluid. Hence we have only a closed system and hence arises the possibility of extending the theory to all kinds of electrical disturbances in spite of our ignorance of the laws of unclosed currents.

The mathematical treatment of this limiting case leads us to Maxwell's equations. We therefore call this treatment a form of Maxwell's theory. The limiting case is so called also by von Helmholtz. But in no sense must this be taken as meaning that the physical ideas on which it is based are Maxwell's ideas.

Figure 15C indicates the state of the space between two electrified plates in accordance with the conceptions of this theory. The distance-forces have become merely nominal. The electricity on the conductors is still present, and is a necessary part of the conception, but its action-at-a-distance is completely neutralized by the opposite electricity of the medium which is displaced towards it. The pressure which this medium exerts, on account of the attraction of its internal electrifications, tends to draw the plates together. In the empty space B there are present only vanishingly small distance-forces.

The fourth standpoint belongs to the pure conception of action through a medium. From this standpoint we ackowledge that the changes in space assumed from the third standpoint are actually present, and that it is by means of them that material bodies act upon one another. But we do not admit that these polarizations are the result of distance-forces; indeed, we altogether deny the existence of these distance-forces; and we discard the electricities from which these forces are supposed to proceed. We now rather regard the polarizations as the only things which are really present; they are the cause of the movements of ponderable bodies, and of all the phenomena which allow of our perceiving changes in these bodies. [See Figure 15D] The explanation of the nature of the polarizations, of their relations and effects, we defer or else seek to find out by mechanical hypotheses; but we decline to recognize in the electricities and distance-forces which have hitherto passed current a satisfactory explanation of these relations and effects. The expressions electricity, magnetism, etc. have no further value for us beyond that of abbreviations.

Figure 15

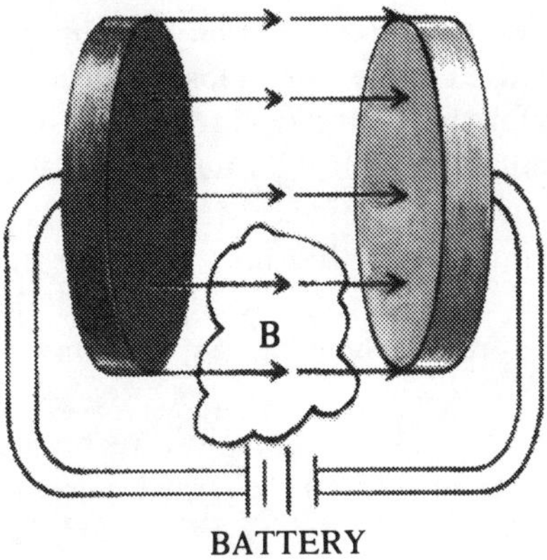

A. Pure action-at-a-distance.

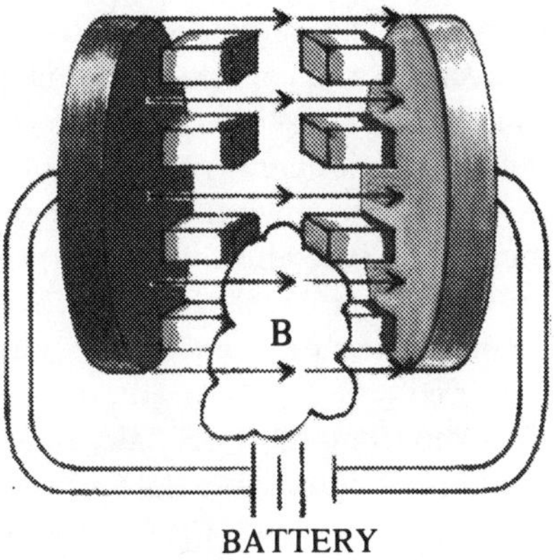

B. The medium between condenser plates is polarized by action-at-a-distance forces.

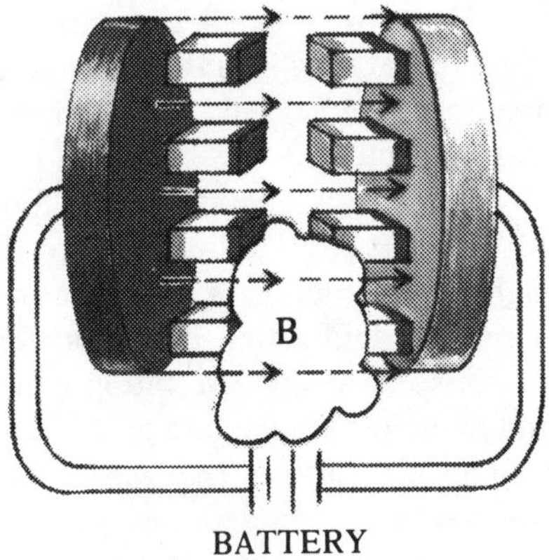

C. The standpoint of modern electromagnetic theory: Action-at-a-distance forces polarize a segment of the medium, giving rise to new action-at-a-distance forces slightly further across the space.

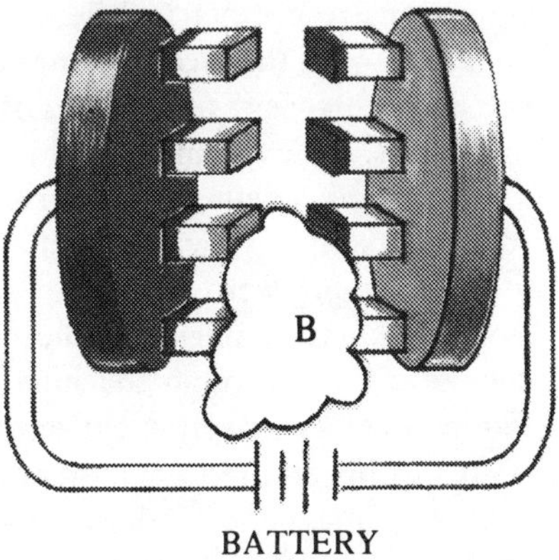

D. The existence of free charges on the condenser plates is denied, with all transmission attributed to polarization of the medium. Note that the direction of polarization is opposite to that of the two previous cases.

Modern physics texts do not, of course, accept the fourth standpoint. The notion of a polarizable aether has been discarded. Modern electron theory reverts to Benjamin Franklin's point of view (also shared by Riemann) that charge will collect on the surface of a condenser plate because of the motion of electrons. Negative, free electricity results from an accumulation of free electrons, positive charge from a loss of electrons. Nonetheless, most texts deceptively refer the geometrical treatment of the field, which is actually based upon Gauss's "flux law" which he conceived as part of the general notion of the curvature of space, to Faraday's far more simplistic and incorrect approximation to that law (Hertz's standpoint number four). Gauss's flux law is taught as the algebra of field theory, the rule that works, but its real geometric implications are obscured.

The Coulomb force law, referred to by Hertz, leads to a vector description of force as a line of force connecting two charged particles. Gauss's law, which is the culmination of potential theory up to that point, establishes transformation of space as a whole. [Figure 16] Force is represented as a potential gradient by streamlines which are perpendicular to lines of equipotential, or contours.

Faraday did not understand the concept of a potential function. He rejected the Coulomb picture of a "central force" which simply connected two charges, but he substituted a pictorial image in its place rather than a comprehensive geometrical description. He approximated the notion of streamlines with the idea of "lines of force." Faraday actually maintained two alternating views of the lines of force, because he could not determine for himself, by experiment, which was correct.

The first is adequately described by Hertz's standpoint number four. From this point of view, he "explained" the curvature of lines of force by the fact that they were produced through the interaction of polarized particles in a quasi-material medium. This interaction had a natural tendency to diffuse through the medium. This is a good description of the production of a convection current through an ionized medium.

It was his failure to demonstrate, to his satisfaction, that there was a polarizable magnetic medium analogous to the electrical one that led Faraday to his alternate concept of lines of force. Now the lines themselves were considered primary.

In his own words:

> I cannot resist throwing forth another view of these phenomena which may possibly be the true one. The lines of magnetic force may perhaps be assumed as in some degree resembling the rays of light, heat, etc.; and may find difficulty in passing through bodies, and so be affected by them, as light is affected. They may for in-

> stance, when a crystalline body is interposed, pass more freely or with less disturbance through it. . . .

Analogously, he considered the curvature of lines of force to result from their tendency to repel each other.

While apologists for Faraday claim that his lines of force are merely a metaphorical expression for vectors, his suggestion that light radiation might be caused by the "plucking" of the lines of force gives to such niceties the lie. It is just this tendency to reductionism, the reification of these *lines* of force, which attracts these same writers. Such verbiage as electrons "twanging" lines of force are not unknown today to monographs in mathematical physics.

Since Faraday did not have the concept of equipotential surface at his command, he imperceptibly shifted his ground over the years until finally, for him, matter became merely an inert receptacle for the imponderable field of his lines of force. His answer to the paradox of comprehending the particle and the force field in the same system banishes the particle. The progression of his ideas is not difficult to follow. Matter, broadly speaking, breaks down into two kinds of substance, those which are primarily dielectric, such as glass and the atmosphere, and those which are conductors, such as metals and salt water. A conductor has a supply of "free electrons" whose response, when placed in an electric field, is to move themselves into an equilibrium position. Hertz's figures best describe the situation, if it is understood that the plates of the condenser are only inactive once they become equipotential surfaces. The only force then exerted is in a direction perpendicular to their surfaces. (The Hertz diagrams of a condenser, by leaving out the wires and battery connecting the plates, can be confusing. The field maintained in those drawings has been "pumped in" by an outside source, perhaps by a battery. Connect those plates with a metal wire and the electrons will migrate to neutralize the field. A current will flow and the condenser will lose its charge.)

Since free electrons in a conductor are mobile, they will only collect at the boundaries of the conducting surface when they are placed in an electric field. Figure 17 shows this schematically.

A perfect conductor does not maintain a field within itself. It is this characteristic which distinquishes a perfect conductor from a dielectric substance. In a dielectric, the electrons can shift their position (become polarized) but they do not migrate. This is shown in Hertz's figures (15A, B, and C), if the space between the conducting plates is visualized as containing an actual dielectric medium such as air, rather than some mythical polarizable aether substance. The dielectric maintains the field throughout its volume. A field will be propagated through both a conductor and a dielectric.

Figure 16

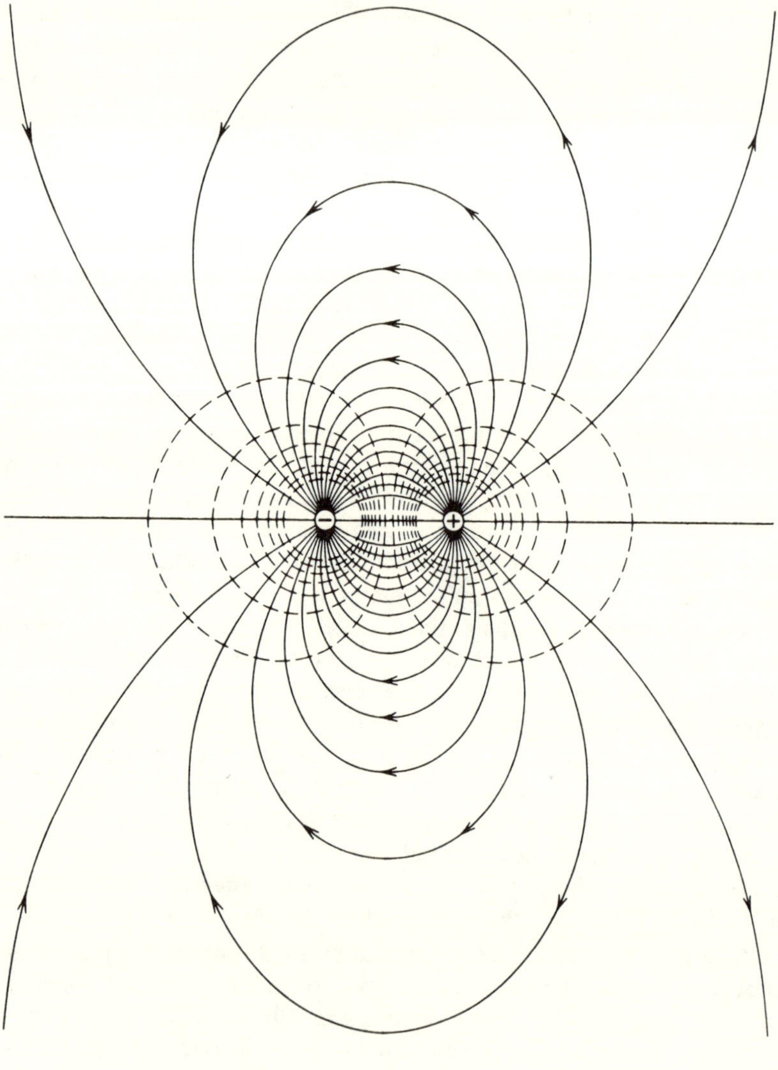

A. Gauss's transformation of space as a whole, in which force lines (solid) are gradients of the potential. Equipotentials are represented as dotted lines.

Figure 16 *Cont.*

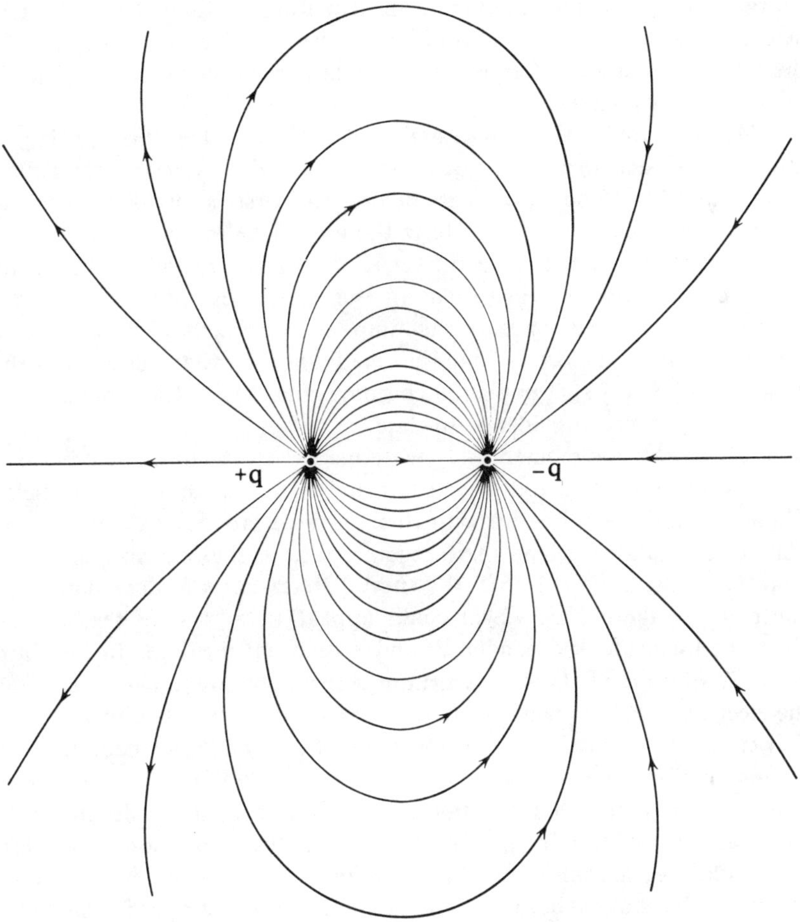

B. In the less sophisticated conception developed by Faraday, the field of two charged objects can be represented by force lines filling all space.

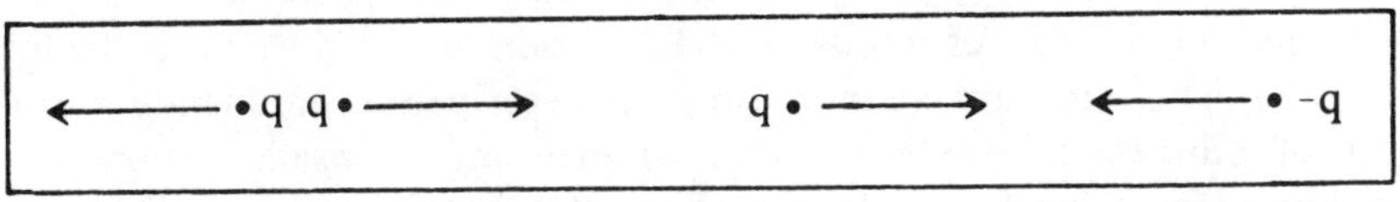

C. The pure Coulomb's Law representation allows for only the force between the objects.

It was Faraday's great achievement to explain the action through dielectrics. While Franklin described the process of conduction of electricity by the flow of charge, Faraday was able to explain the induction of charge through the polarization of an insulating medium. Unfortunately he exaggerated the implications of his discovery to develop a "pure" field theory which treated matter as a mere container for the field. The point is underscored by Hertz.

Without doubt, theories of action-at-a-distance beg the question of how the field structures itself; nonetheless, Hertz's "solution" has all the flaws located by Zeno 2,500 years before. But first, let us dispose of the most glaring weakness of the Hertz-Faraday field theory. According to Hertz, empty space acts as a dielectric. A condenser will maintain the charge on its plates even in a vacuum. But, that does not justify the converse assumption that space, or the "imponderable aether," has the same properties as a dielectric material medium. In what sense can the molecules of empty space be said to become polarized? Clearly none in the face of all empirical evidence to date.

Assume for the moment that an aether exists which can be described as a sequence of polarized molecules and that this aether is the field. These molecules will line up with their positive face toward the positive plate of the condenser and their negative face in the opposite direction. *Nota bene,* this is Figure 15D, the special Hertz-Faraday explanation, in contrast to Figure 15C, which could appear in a modern textbook to show the behavior of a condenser and a material medium. In the third interpretation, the field is shown running from positive to negative, while the electrons strain to travel from the negative to the positive plate. Since they cannot move because of the structure of the insulating medium, they are merely displaced or polarized. In Figure 15D, however, the absence of field lines, as well as the direction of polarization of the molecules indicates the changed standpoint. The field is not other than these molecules. The molecules are the quasi-material field. The condenser plate only appears to be charged because it provides a bounding surface for the field.

It is the implied epistemological fallacy that is of greater intrinsic interest. Divide the aether as finely as you like. How is the action propagated from one molecule to the next? One either resorts to action-at-a-distance in the small or faces all the consequences of Zeno's paradox.

The answer lies, of course, in the direction to which we have already pointed. The infinite is no more properly described as an unending process of infinitesimal subdivision than by unending extension. Action-at-a-distance was ackowledged as a convenient subterfuge even by Newton. With this admitted elision of the inbetweeness of propagation and the accompanying assignment of an infinite velocity to the speed of the electric field, it was possible to make the first approximations to field theory.

Figure 17

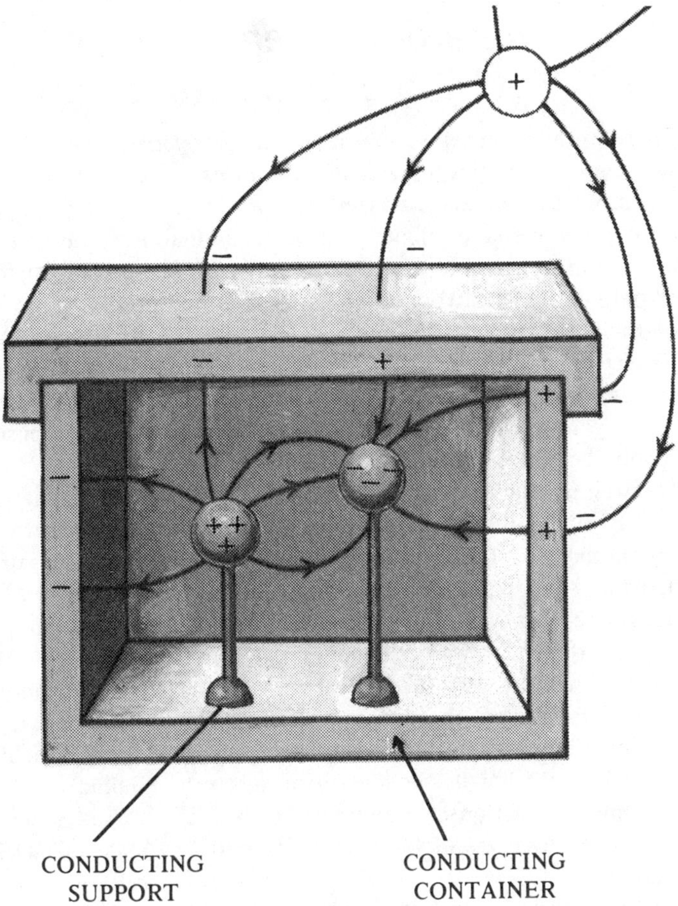

By bringing static charges near to a conducting box, a polarization can be induced on it. Charges reside only on the surfaces of the box, not in the metal. Also, there is no net charge accumulated on the box. Polarization occurs due to the attraction and repulsion between charges. Mobile charges in the conductor will seek an arrangement where they can come to rest. An unpaired charge in the material will be attracted to one surface and repelled from the other by any charge already built up on the surfaces.

Gauss, for example, was very explicit on this point in his letters. It was Riemann who first advanced beyond this point with his theory of the "retarded" potential.

Retarded Potential vs. Action-at-a-Distance

There were two crucial hypotheses which governed scientific work in the late eighteenth and nineteenth centuries. The first was that the physical universe was characterized by lawfulness so that the apparently disparate phenomena of light, electricity, magnetism, and gravitation should be comprehensible as a unified process. The second was that the unified field of the first hypothesis was a subsumed predicate of the evolution of the universe.

Expectation as to the character of universal lawfulness reflected the otherwise philosophical bent of the experiment. The French materialists and Faraday, as well, understood this lawfulness as a superimposed order grafted on the universe.

Maxwell accepted this view, but he was most subject to the insidious effects of British empiricism and almost behind his own back was able to hold competing, contradictory assumptions at one and the same time. Under the guise of explaining Faraday's "pure" field theory in mathematical language, Maxwell introduced concepts derivable from potential theory which depended upon the generative role of the particle. These, in fact, cohered with his own theoretical work in statistical thermodynamics which were based on the theory that heat is generated by kinetic molecular action. The French materialists also accepted this particle-field dichotomy, which could hardly be said to be decently cloaked by the thinly veiled disguise of action-at-a-distance. It is their immodest nakedness which is a positive quality when compared to Maxwell's Victorian obscurantism.

Hertz bitterly complained about the theoretical confusion in Maxwell's classical text, *Electricity and Magnetism*, going so far as to say that "Maxwell's theory is Maxwell's system of equations. Every theory which leads to the same system of equations and therefore comprises the same possible phenomena, I would consider as being a form of special case of Maxwell's theory." Such a situation is plainly a mess, as Hertz is at pains to point out, but modern texts frequently quote his remark in support of their own disregard of theory. In such circumstances, theory becomes little more than a mnemonic device to aid the memory in retaining disparate sets of rules. Hertz, who was more rigorous than Maxwell in practice, ironically subscribed to this debasement of theory in his own commentaries on the epistemology of science.

The philosophic assumptions which underlie the second hypothesis were shared by Riemann, Cantor, Hegel, Marx, Benjamin Franklin and the other leaders of the American Revolution, and, one can suspect, that secret negentrope Gauss, who, after all, immediately endorsed Riemann's presentation of the basis for a negentropic universal geometry and is known to have suppressed his own most revolutionary mathematical discoveries. The universe is a self-contained, self-developing, self-expanding continuum. It was Riemann who saw that this process is properly understood as a geometry.

Any given geometry is valid by approximation only within a given limit of "convergence." Beyond that circle of convergence, small-order differences begin to accumulate in such a way as to render any decisions made on the basis of the geometry unpredictably false. Beyond the circle of convergence a new geometry is necessary. It is at the bounding point, between two such geometries that singularities appear. These singularities, by appearing in both geometries, provide the basis for understanding the causal geometry which is the driving force for change from one geometry to the next.

Riemann's life work was to give expression to this humanist, Cartesian understanding of natural law in the language of geometry. This was not mere exercise in translation or appeal to descriptive analogy. The conceptual framework of Riemannian geometry and its amplification by Cantor are the precise basis for rigorous further accomplishment in every area of science, under which art, history, and political economy are properly subsumed as particular sciences.

It is appropriate here to quote at length from Riemann's philosophical notes, published posthumously.

Antinomies

Thesis	Antithesis
Finite things, imaginable things.	Infinite things, conceptual systems which lie on the borders of the imaginable.
Finite time and space elements.	Continous things.
Freedom, i.e. not the ability to initiate things in an absolute sense, but to decide between two or more given possibilities. In order that decision through	Determinism. No one engaged in action can give up the conviction that the future will be codetermined through his actions.

free will be possible, in spite of the totally determinate laws of the action of the conceptions, one must assume that the psychic mechanism itself has or at least, in the course of its development, assumes the characteristic quality of bringing about the necessity of the latter.	
A God active in time (world government).	A timeless, personal, all-knowing, all-powerful, all-good God (providence).
Immortality. Freedom is entirely compatible with the strict lawfulness of the course of nature, but the concept of a timeless God is not tenable alongside of it. Rather, the restriction which omnipotence and omniscience suffer through the freedom of the creatures, in the sense presented above, must be removed through the acceptance of a God active in time, of a guide for the hearts and fates of men. The concept of providence must be expanded, and be partially replaced by the concept of world government.	A thing in itself which is the basis of our transient phenomenal existence and is equipped with transcendental freedom, radical evilness, and an intellectual character.

Riemann sought to test his hypothesis by recasting field theory. Action-at-a-distance is an abstraction from time and obviates any possibility of development. Practically, therefore, he developed the notion of a retarded potential, which travels at a finite velocity, to replace the notion of a static potential which is put into place at an infinite velocity. With the retarded potential, he was able to directly recast the potential function as a volume integral

$$\frac{1}{4\pi}\int\frac{\varrho dv}{r}$$

whose solution in points outside the charge will be the Laplacian

$$\frac{\partial^2 V}{\partial x^2}+\frac{\partial^2 V}{\partial y^2}+\frac{\partial^2 V}{\partial z^2}=\frac{\partial^2(1/r)}{\partial x^2}+\frac{\partial^2(1/r)}{\partial y^2}+\frac{\partial^2(1/r)}{\partial z^2}=0.$$

If it is necessary to discuss a potential function which varies in space with time, the potential is no longer a function of the charge divided by the distance r, but the function must express the fact that electromagnetic effects are propagated with the velocity c. A particular motion of charge at a particular time produces potentials and fields at a distance r at time r/c later. The retarded potential function is

$$\frac{1}{4\pi}\int\frac{\varrho[t-(r/c)]}{r}\,dv=V$$

which gives rise to

$$\frac{\partial^2 V}{\partial x^2}+\frac{\partial^2 V}{\partial y^2}+\frac{\partial^2 V}{\partial z^2}=\frac{1}{c^2}\frac{\partial^2 V}{\partial t^2},$$

an equation describing a potential wave. If r runs parallel to the x-axis, the validity of the formula can be easily understood. Take any function of the form $y=f[t-(x/c)]$; then first and second derivatives with respect to both x and t yield

$$\frac{\partial y}{\partial x}=\frac{\partial f[t-(x/c)]}{\partial t},\qquad \frac{\partial y}{\partial t}=-\frac{1}{c}\,\frac{\partial f[t-(x/c)]}{\partial[t-(x/c)]},$$

$$\frac{\partial^2 y}{\partial t^2}=\frac{\partial^2 f[t-(x/c)]}{\partial[t-(x/c)]^2},\qquad \frac{\partial^2 y}{\partial x^2}=+\frac{1}{c^2}\,\frac{\partial^2 f[t-(x/c)]}{\partial[t-(x/c)]^2}.$$

Clearly, then,

$$\frac{\partial^2 y}{\partial x^2}=\frac{1}{c^2}\,\frac{\partial^2 y}{\partial t^2}.$$

which expresses the fact that the electrical field is propagated at the speed of light by a pulse or wave. To maintain a constant argument, t-(x/c), for the function corresponding to a particular point on the pulse, the displacement x must increase with time t. Consequently, a designated point on the pulse travels in the positive x direction at speed c. The wave may be made up of one pulse, or a succession of pulses of different magnitudes, depending upon the activity of the field. A static field can be considered as an almost square pulse of infinite length traveling outward. [Figure 18]

This equation for the retarded potential and an expression for the retarded "vector" potential A, which we have not yet discussed, together lay a perfectly adequate basis for the "Maxwellian" wave equations. We shall return to this at greater length.

Figure 18

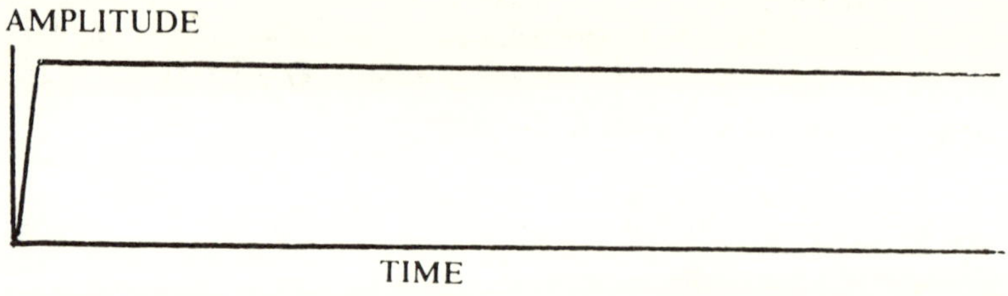

A. Time evolution of a square pulse representing a static field. After the initial rise passes a point in space, no further variation is observed at that point.

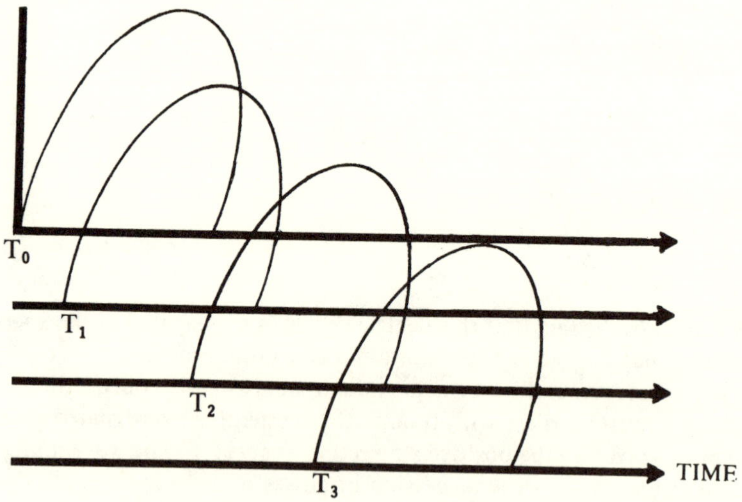

B. A disturbance, such as a single shake of a rope will pass along the rope and leave each point in an undisturbed condition after the pulse passes by.

Figure 18 Cont.

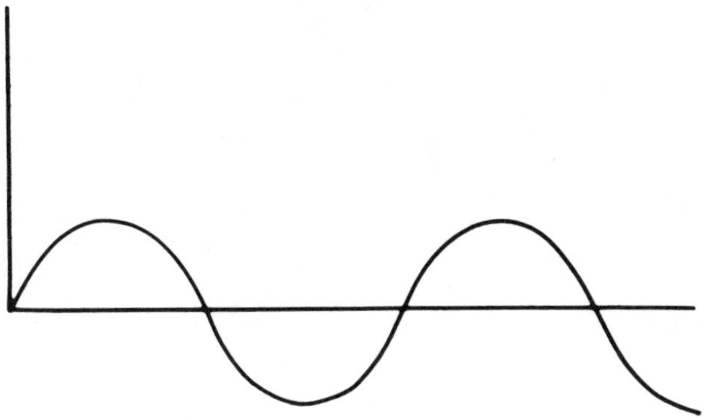

C. The sine (or cosine) function is a useful representation for a regularly repeating wave disturbance.

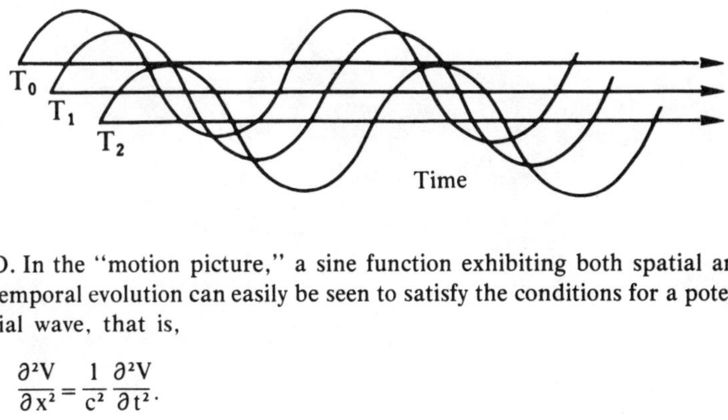

D. In the "motion picture," a sine function exhibiting both spatial and temporal evolution can easily be seen to satisfy the conditions for a potential wave, that is,

$$\frac{\partial^2 V}{\partial x^2} = \frac{1}{c^2}\frac{\partial^2 V}{\partial t^2}.$$

Let $V = V_0 \sin(t - x/c)$, then

$$\frac{\partial V}{\partial x} = -\frac{V_0}{c}\cos(t - x/c), \quad \frac{\partial V}{\partial t} = V_0 \cos(t - x/c),$$

$$\frac{\partial^2 V}{\partial x^2} = -\frac{V_0}{c^2}\sin(t - x/c), \quad \frac{\partial^2 V}{\partial t^2} = -V_0 \sin(t - x/c).$$

There are, of course, many other functions such as $V = V_0(t - x/c)^2$, which also satisfy the conditions for a potential wave, so that boundary conditions appropriate to a particular physical situation must be applied in order to arrive at the correct function for that situation.

CHAPTER IV

Geometrical Physics

The last great mathematical physicist was Felix Klein. The extent to which his time was devoted to salvaging the accomplishments of Riemann, his predecessor at Göttingen University, interpreting them, and passing them on to the German school system is a testament not to his own lack of creative scope, but to the vicious decline in mathematical physics in this century. He was able to partially stem the flood tide toward abstract formalism in "pure" mathematics and indifferentism in physics. Even today, the quality of science education in Germany far surpasses that in the United States, because it remains grounded in the pedagogical program of Felix Klein.

The point of view, which we quote here, aptly summarizes Riemann's own intuition about his own work.

> Many have thought that one could, or that one indeed must, teach all mathematics deductively *throughout*, by starting with a definite number of axioms and deducing everything from these by means of logic. This method, which some seek to maintain upon the authority of Euclid, certainly does not correspond to the historical development of mathematics. In fact, mathematics has grown like a tree, which does not start at its tiniest rootlets and grow merely upward, but rather sends its roots deeper and deeper at the same time and rate that its branches and leaves are spreading upward. Just so — if we may drop the figure of speech — mathematics began its development from a certain standpoint corresponding to normal human understanding and has progressed, from that point, according to the demands of science itself and of the then-prevailing interests, now in the one direction toward new knowledge, now in the other through the study of fundamental principles. For example, our standpoint today with regard to foundations is different from that of the investigators of a few decades ago; and what we today would state as ultimate

> principles will certainly be outstripped after a time, in that the latest truths will be still more meticulously analyzed and referred back to something still more general. *We see, then, that as regards the fundamental investigations in mathematics, there is no final ending, and therefore, on the other hand, no first beginning which could offer an absolute basis for instruction.*

These words were written by Felix Klein in 1908 to introduce his work *Elementary Mathematics From an Advanced Standpoint.* They are just as appropriate today.

Riemann's life work was a polemic against the Lagrangian universe, but it was very much determined by the context of that universe whose overthrow he accomplished. Not only is his work necessary to any competent treatment of field theory, but it is the physical intuitions derived from the criticism of existing field theory which informed his work and offered the launching pad for his more theoretical accomplishments. In that sense, the intuitions of potential theory were the basis for his own theoretical accomplishment as a counterpoint.

While, at first glance, Riemannian geometry as applied to electromagnetic field theory appears to merely offer the opportunity for greater rigor in the treatment of singularities and the possibility of determining which of the "inessential" singularities within the particle-field duality may be brushed aside as rigorously susceptible to linear approximation, this is only a first impression.

The distinction between nonessential, "polar" singularities, which can be linearly approximated to any desired degree of precision, and essential singularities, which cannot be so approximated, is not the decisive issue. The particle may be treated as a polar singularity within the confines of classical electromagnetic theory, yet this is simply to ignore those essential singularities overlooked by electromagnetic field theory as such, but which define its internal geometry. Conversely, the lawful ordering of essential singularities only appears insoluble at any given level of society. It is not the identification and classification of singularities which is essential to Riemann's work, it is the development of a notion of successive, higher-order manifolds whose geometry is determined by the interplay of these singularities to create increasingly complex topologies.

By the time he was 21 years old, Riemann had already laid the foundations of his major accomplishments. The process of internal revolution that he went through must have been very much like the experience of Karl Marx in 1844 when he wrote his germinal *Manuscripts.* Here was Riemann, the devout son of a Lutheran pastor, bursting to give human expression to his active God, filled to overflow with new ideas, each one shattering the bounds of a static universe.

We quote from two letters to his father written in 1853 when he was 27 years old. He was just at the point of giving the lecture concerning the foundations of geometry from which we have quoted at length. He was obliged to prepare the paper for presentation at a trial lecture before the Göttingen mathematics faculty as a precondition for appointment as lecturer. The topic was chosen by Gauss from the three submitted by Riemann. To his father he wrote:

> So I am again in a quandary, since I have to work out this one. I have resumed my investigations of the connection between electricity, magnetism, light, and gravitation and I have progressed so far that I can publish it without a qualm. I have become more and more convinced that Gauss has worked on this subject for years and has talked to some friends (Weber among others) about it. I tell you this in confidence, lest I be thought arrogant — I hope it is not yet too late for me and that I shall gain recognition as an independent investigator.

Some months later, he again writes his father on the subject:

> I became so absorbed in my investigation of the unity of all physical laws that when the subject of the trial lecture was given me, I could not tear myself away from my research ... Having finished two weeks after Easter a piece of work I could not get out of, I began working on my trial lecture and finished it about Pentecost.

The lecture took him only about five weeks to complete. At the same time, he prepared the article, "On The Laws of Distribution of Potential Energy..." His letter continues:

> Weber and some of his collaborators in the electromagnetism laboratory had made very exact measurements of a phenomenon which up till then had never been investigated, the residual charge in a Leyden jar. ... I sent him my theory of this phenomenon, having worked it out specially for his purposes. I had found the explanation of the phenomenon through my general investigations of the connection between electricity, light and magnetism. ... This matter was important to me, because it was the first time I could apply my work to a phenomenon still unknown, and I hope that the publication will contribute to a favorable reception of my larger work.

Riemann discovered the concept of negentropy for himself when he was 21 years old. At that time, he developed the fundamental topological concept of those paths through space which define reproductive cycles of increasingly higher-ordered complexity. The Cauchy-Riemann equations, which allow the integration of harmonic functions, are a first approximation of this idea. A simple example is a space with one polar singularity. The integral which sums the path around such a singularity in three dimensions will have the value or strength $4\pi\varrho$ when a complete cycle is made around it, which is equivalent to the residue theorem for integration in the complex plane.

At the same time as Riemann concerned himself with the geometry of space in the small, he was developing the notion of an immanent metric which could describe the process of self-development of the universe as a whole. In this conception, the notion of curvature was used as a first approximation to the successively complex topologies which characterize the description of potential fields when these are viewed from a macroscopic universal standpoint.

In both the macrocosm and the microcosm, the singularity defines the boundary line between two geometries. Field theory in microcosm deals with singularities as defining the inside and outside of space, bounding related, but separate geometries. Macrocosmic field theory looks at these boundaries not as existents, but as the forcing moments which determine a process of development which subsumes and supersedes them. Thus, macrocosmic field theory, the theory of the developing curvature of space, is always necessarily at variance with any local measure of the field which accepts the given singularities as existents in themselves and to which the laws of the universe as they are reflected by science must accomodate themselves. Riemann states the problem somewhat differently in the *Hypotheses.* Here he locates invariance in the macroscopic astronomical universe. The contrast between the macroscopic and microscopic universe is, however, sufficiently clear to establish our point.

> If one premises that bodies exist independently of position, then the measure of curvature is everywhere constant; then from astronomical measurements it follows that it cannot differ from zero; at any rate its reciprocal value would have to be a surface in comparison with which the region accessible to our telescopes would vanish. If, however, bodies have no such nondependence upon position, then one cannot conclude to relations of measure in the indefinitely small from those in the large. In that case, the curvature can have at every point arbitrary values in three directions, provided only the total curvature of every portion of space be not appreciably different from zero. Even greater complications may arise in case the line element is not representable, as has been premised, by the square root of a differential expression

of the second degree. Now, however, the empirical notions on which spatial measurements are based appear to lose their validity when applied to the indefinitely small, namely the concept of a fixed body and that of a light-ray; accordingly it is entirely conceivable that in the indefinitely small the spatial relations of size are not in accord with the postulates of geometry and one would indeed be forced to this assumption as soon as it would permit a simpler explanation of the phenomena.

As a subsumed feature of his thought, Riemann developed a first geometric approximation to the development of a macroscopic immanent metric customarily known as the Riemannian curvature tensor, and a notion of transformations of space which maintain topological invariance known as the theory of Riemannian surfaces.

The Complex Number Plane

To understand these ideas, it is necessary first to understand the complex number plane. Gauss is credited with its discovery, although Euler and others before him had used it as well. As early as 1545, the algebraic efficacy of complex numbers was readily admitted. The following problem was posed by Cardano at that time. Divide the number 10 in two parts so that their product equals 40; or, in modern notation, solve for x when $x^2-10x+40=0$. There are two "numbers" $5+\sqrt{-15}$ and $5-\sqrt{-15}$ (i.e., 25+15=40) which are equal to x according to the well known formula

$$x=\frac{-b\pm\sqrt{b^2-4ac}}{2a}.$$

As Cardano observed, they offer a perfectly adequate solution to the problem, except, of course, that the square root of a negative number was not considered to be a number. The usefulness of this kind of formal solution led to the adoption of these quasi-numbers by a simple extension of the rules of calculation. The square root of negative one became known as i (Euler's term, a carry-over from the idea that these useful numbers were not quite real — but "imaginary.") The rules of their calculation follow logically once it is established that the square of a number such as $5i(\sqrt{-5})(\sqrt{-5})$ is -25. The product of negative and positive 5i is positive 25. (The numbers 5+5i and 5-5i are known as conjugates.)

Operations with these numbers follow the same logical rules as operations with ordinary numbers. An example is the sum of 5+6i and 3+7i, which is equal to 8+13i; therefore a complex number can also be written simply as a number pair. In this case, we have (5, 6) and (8, 13). The first number is always the real component and the second the imaginary. It is on this slender basis that it is customary to motivate Gauss's use of the number plane. If complex numbers can be represented by number pairs, then it follows that they can be treated graphically if we simply make the y-axis an axis of imaginary numbers. [Figure 19] However, the explanation overlooks the exciting difference between the complex number plane and the ordinary plane. Even within the domain of linear transformations, this is clear. The algebra of complex numbers is a complete direct expression for the geometry of linear transformations (translations, rotations, and stretchings). Again it will be simplest to explain by example. The product of 3i and 4 is 12i. We have stretched 3i by a factor of 4. The product of negative 3i and 4 is negative 12i, again a familiar type of result. The product of negative 3i and negative 4, which is positive 12i, may have confused you at school because it implies a 180 degree rotation. This is usually overlooked to emphasize the convenience of the result which is usually ultimately convincing. Now, however, if we multiply by 4i, we see that rotation is indeed a fundamental arithmetic operation. The product of 3i and 4i is negative 12, a stretching and a 90 degree rotation. [Figure 20]

It is but a short step to represent these numbers by trigonometric relations. Let's take the number $12+12i$. That number can be represented by a vector which makes an angle of 45 degrees with the x-axis. Its length is $12\sqrt{2}$. We can represent its real components as the product of $12\sqrt{2}$ and the cosine of 45 degrees, $1/\sqrt{2}$, which, of course, is 12. Similarly we can denote its imaginary component by the product of $12\sqrt{2}$ and the sine of 45 degrees, again, in this case, $1/\sqrt{2}$, times i, or 12i. In other words, we can represent any complex number as $r(\cos\Theta+i\sin\Theta)$, where r is the length of its vector or its absolute value, and the angle Θ is determined as above. Now we are in a position to apply a convenient theorem from trigonometry which allows us to multiply two numbers: if A equals $r_1(\cos\Theta_1+i\sin\Theta_1)$ and B equals $r_2(\cos\Theta_2+i\sin\Theta_2)$ then the product AB equals the product $r_1r_2[\cos(\Theta_1+\Theta_2)+i\sin(\Theta_1+\Theta_2)]$. Multiplication may be represented directly without recourse to trigonometry, but the use of trigonometry shows the result most clearly. Complex number multiplication is a combination of two processes. The original number-vector, the absolute value of A, will be stretched by the absolute value of B and it will be rotated by an amount equal to the angle which B itself makes with the x-axis. [Figure 21] Addition and subtraction follow the ordinary parallelogram law of vector addition [Figure 19] and division is the inverse of multiplication.

So far, we have just assembled the tools which will enable us to explain Riemann's geometric conceptions. With one further addition, we

Figure 19

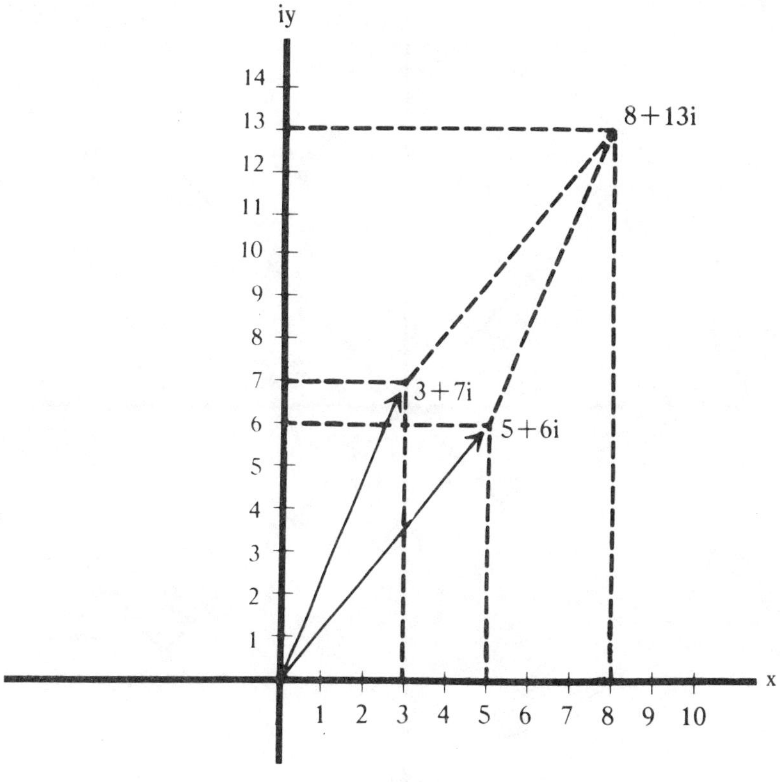

Geometrical representation on a complex plane of the rule for adding complex numbers. Note that this is identical to the parallelogram rule for addition of vectors.

---

will be in a position to begin to apply these tools to the study of field theory. To Gauss's complex number plane, Riemann appended the "ideal" point at infinity. In order to do this, he applied the method of stereographic projection familiar in two-dimensional mappings of the earth's surface which preserve direction, but not magnitude. (Such maps will generally represent polar masses such as Greenland as disproportionately great.)

Figure 20

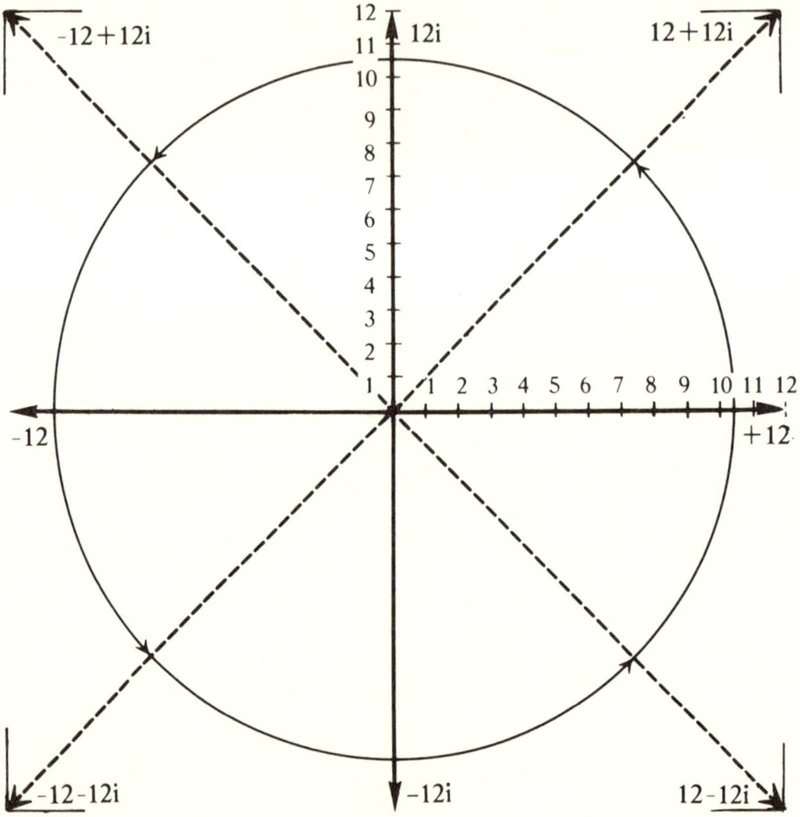

Multiplication of the complex number a+ib by i corresponds to a counterclockwise rotation through 90 degrees in a complex plane. The reader can check this result for the number 12+12i.

On the complex number plane, we place a sphere of diameter 1 in such a manner that it touches the plane at the origin of coordinates, 0. So that we may be able to express ourselves conveniently, we make use of the usual geographical terminology on the sphere, and accordingly call the point of contact, 0, the south pole, and the diametrically opposite point, the north pole, N. We now consider the rays, issuing from this north pole, which intersect the plane, and consequently also intersect the sphere at a

Figure 21

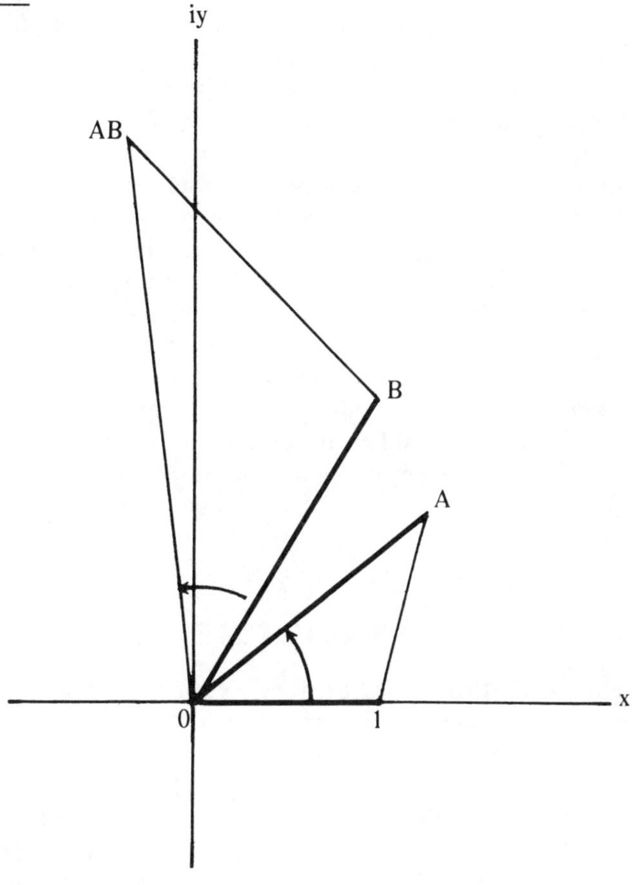

Multiplication of two complex numbers A and B represented trigonometrically. Plot the points 0, 1, A, and B. On the segment 0B, construct a triangle similar to 01A in such a manner that 01 and 0B are corresponding sides. The third vertex of this triangle becomes the point AB.

second point (distinct from N). We associate the point (distinct from N) on the sphere with that point of the plane which lies on the same ray. Obviously, to every point P of the plane, there corresponds, in virtue of this association, precisely one point P′ of the sphere, distinct from N, and conversely. The surface of the sphere (from which one must imagine the north pole to be deleted) is mapped in a one-to-one manner on the plane. The north pole will represent the point at infinity. It is clear that the parts of the sphere lying in the neighborhood of the south pole are only slightly distorted, while those situated near the north pole undergo a violent distortion. Every meridian on the sphere is mapped onto a straight line in the plane; every line of latitude on the sphere is mapped onto a circle in the plane; any two circles or curves in general on the sphere intersect at the same angle as their images in the plane. [Figure 22]

The inclusion of the point at infinity is not merely a convenience which allows us to achieve a consistent mathematical treatment of the point at infinity, but, by relegating the unending bad infinity of mere extension to the same status as every other point, it forces a proper notion of true infinity. The treatment of the ideal points as numbers follows directly from considerations of projective geometry, which we shall deal with next. But, first, the importance of the concept of the stereographic mapping cannot be overlooked. The apparently infinite, the unending plane, becomes the finite, but unbounded — the sphere. What appears on the plane to be the potential for unending simple extension (line) is shown for what it must be, the repetition of cycles (meridian). Must such cycles be simply repetitive? Not at all if, for example, the sphere is expanding during the process.

Curvature

A more sophisticated measure for the geometry of the sphere taken as a universe is the notion of curvature. The smaller a sphere, the greater its curvature; the greater the radius of a sphere, the closer it approximates to a plane by flattening out. The curvature of a plane is zero. Gauss was the first person to describe the curvature of a surface. His method was to take all the normals (the unit vectors perpendicular to the surface) to a surface and place them so that their points of contact to the surface came together. [Figure 23] These points of contact would then become the origin of a unit sphere. Their tops would mark off an area on the surface of the sphere. The limit, as the area of original surface is reduced to a point, of the ratio of the area marked off on the sphere to the area on the original surface is Gauss's measure of curvature. Gauss, of course, devoted himself to ways of finding that measure which would be less cumbersome. One such method is to grid the sphere and compare the distance between two points on a longitude when one is constrained to walk on the sphere, rather than traveling straight through space.

It is equally possible to measure the curvature by measuring the projection on the complex plane between two equal longitudinal arcs on the surface of the sphere. The smaller the sphere becomes, thus increasing its

Figure 22

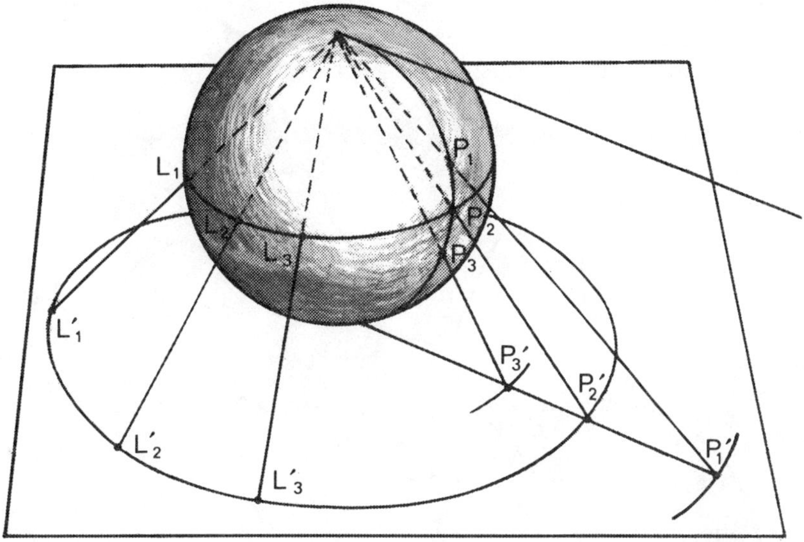

Lines extended from the north pole through a meridian project onto the plane so that the closer two equally spaced points on the meridian are to the north pole, the farther apart they will appear in the projection. Lines of latitude project as concentric circles of increasing radius going from the south to the north pole. The meridians intersect these circles as radii.

curvature, the greater the discrepancies in the mapping of magnitude on the plane. It would not be necessary to take the original Euclidian measure of distance as a guide; we could just as well measure successive changes in the projection itself from one period to the next. It should not trouble us that our imaginary sphere must become smaller in order to indicate a higher order of potentiation or curvature, as it is often useful to invert measure relations in a transformation in order to exemplify certain invariant features.

It is easy to extend the qualitative notion of curvature to the potential field. Consider that "distance" is measured by the time it takes to travel equivalent paths with the same expenditure of work, or conversely one can equalize the time and measure the different amounts of work necessary to accomplish the trip. Clearly the "spherical distance" will vary according

Figure 23

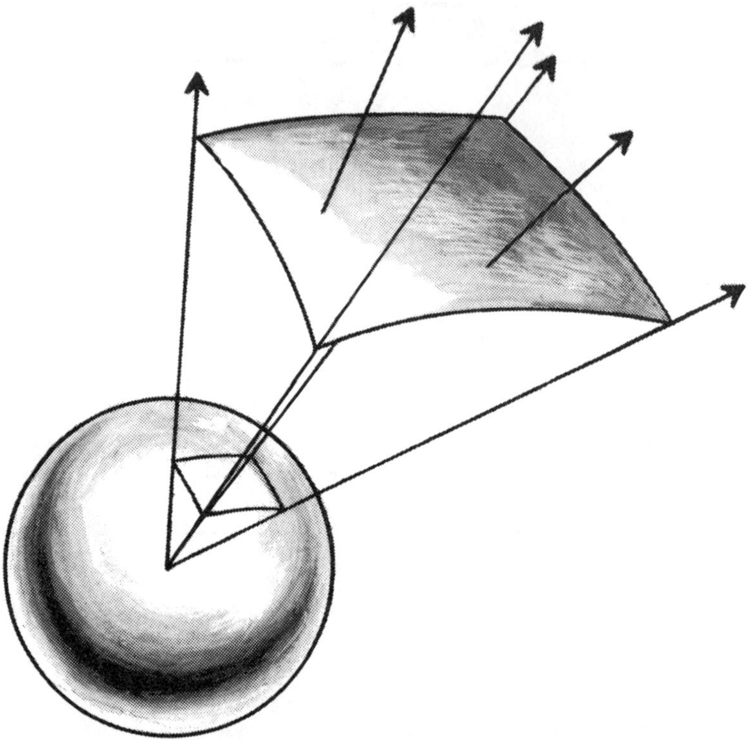

To illustrate Gauss's method of determining curvature, imagine the directions of all the normals to the rectangular surface of the small sphere. Compare these directions to those on the extension of the rectangular surface onto a larger sphere. When all the points of contact are brought together, the area marked off on the unit sphere will be the same for both cases. Consequently, the ratio of this area to the area of the original surface will be larger for the smaller sphere. The curvature is said to be greater for the smaller sphere.

---

to the gradient of the field. If we have a field where energy is conserved, we will find that the "curvature" is zero. (Consider the case of walking up and then down a hill. If energy is conserved, the "curvature" of the field is zero, despite the geometrical ups and downs.)

It was this notion that Riemann used to develop a first linear approximation to curvature in a space free of singularities. As such a heuristic approximation, it is a useful image upon which to base the concept of a negentropic process of developing curvature. To use Riemann's curvature tensor as he developed it by first approximation formulae as a basis for a unified field theory was Einstein's pathetic lunacy.

Only in Euclidean geometry is it possible to transfer a vector from one point to the next without regard to what path is chosen. Consider the vector as a ruler; then, with the ruler and a compass, we can measure the distance between two points with appropriate corrections for those instances when we may have veered from a straight line. The straight line is the appropriate measure of distance, but, whatever path we take, the ruler-vector will still remain an adequate measuring instrument. Space is flat in Euclidean geometry. Now consider a non-Euclidean geometry. How, for example, would you compare the area of Greenland to that of New York City. Obviously, it is necessary, figuratively, to use two different, but related rulers. The relationship of the rulers can then be used to determine curvature.

When a vector X, defined as a function of its position, undergoes a parallel displacement along a closed curve, upon its return to its point of departure, it may have changed its direction. We can say that it has become a new vector X*. The difference then between X and X* can be used to calculate the curvature tensor (that is, the measure of curvature). At any given point, if there is no essential singularity, we can hope to construct a coordinate system (like a grid of "rulers") such that the rate of change of a vector traveling on it will be a minimum, $dX/dt=0$. We now seek a series of coefficients to compensate for the degree to which the coordinate systems of each point change. Using these coefficients, we can compute the curvature tensor.

This, however, excludes those most interesting singular points in which linear approximation is impossible — where the vector dX/dt neither exists nor can be approximated. In essence, we are back in the field from which the particle must be removed if we are to determine its measure. If we accept the particle as a black-box field container, which then maps it to the paradoxical domain of the Copenhagen School, in other words, if we accept Gauss's flux law as a valid approximation for field strength, then the singularity at the boundary can be linearly approximated. Such nonessential singularities are known as poles.

While an essential singularity will never be describable by linear approximations, it is useful to refer briefly here to Georg Cantor's treatment of the transfinite. The concept which we want to express is the following: Historical evolution constantly poses new essential singularities to the human species as problems which must be solved if the species is to survive. These singularities define the nexus of two successive geometries which must be bridged by a nonlinear creative subsumption of the small-order processes in the first which can be developed as a basis for a higher-order potentiation in the second. This having occurred, the "solution" exists as a repeatable social fact. For example, the process of introducing fusion technology will take place in successive waves on an increasingly broad scale. This will allow debugging of the initial stages of transformation, but will be essentially repetitive, assuming the amount of planning which will have been necessary to take the first giant steps.

The "linear" extension of any given technology can be represented by Cantor's step-wise ordering within the transfinite. Every succeeding transfinite cardinal number subsumes those numbers which precede it. Evolution is telescoped within present practice. From this standpoint, essential singularities can be placed in two categories: those which await solution and those which have been subsumed within existing practice. Poles or nonessential singularities can then be further generalized to express differences in ordinality within a given cardinal number. Such, for example, would be the subdivision of the particle-field collectivity to express the boundary between the interior of the particle and that portion of the field exterior to it within the present cardinality of prefusion technology, which leaves certain questions necessarily open if descriptively approachable. (A full treatment of Cantor's development of transfinite numbers is presented in Parpart, Uwe. "The Concept of the Transfinite." *The Campaigner.*)

Unless the given surface is flat, the vector X' will change direction as it completes any circuit to become the new vector X^*_i. The curvature tensor which we developed by considerations of the constancy of length can also be measured by the difference in direction between the two vectors divided by the area of the circuit which they traveled as that area approaches the limiting value zero.

This is an application of Stokes's formula, discovered independently by Riemann. In electromagnetic theory, it is used to assert that we can measure the work accomplished or used up in any given circuit either by a line integral, which takes the force around the circuit, or by an area integral, which sums up the rotation over the area covered by the circuit. In this case, Riemann has used the line integral to arrive at the same measure of curvature as Gauss, who considered the surface as a whole. On a two-dimensional surface, the measure will be the same except for the factor $-\frac{3}{4}$. The measure is determined by consideration of the difference in the second order variation of the rulers according to the variation of path.

The fundamental idea which takes Riemann beyond Gauss is the assertion that the length of the vector need not depend upon any original Euclidean measure, but can be developed as an implicit standard of measurement. Therefore, the given surface need not be embedded in Euclidean space. We now have a heurism to express the principle of ordering by which higher-ordered manifolds are successively developed. The same sort of problem arises within the domain of political economy when we wish to assign a proper value to commodities in comparison with their particular historical values which may be distorted for any number of reasons relating to the uses of political power or the subsequent development of technology. In this case, the metric by which price must be adjusted to guarantee a proper reproductive "constancy" will be determined by the necessity to increase the future productive potential of labor — its labor power. The value of any object is not a linear function of its own past performance, nor even its future role in production. It is a function of the entire field which subsumes then and now.

While Riemann extended Gauss's work on curvature in this way to lay a firm basis for a multidimensional non-Euclidean geometry, he was concerned with the problem of essential singularities, those "infinite conceptual systems which lie at the borders of the representable." He, therefore, devoted most of his effort to the broader questions of projective transformations opened up by his development of complex function theory.

General Transformations

We have looked at the projective relationship which allows points on the complex plane to be represented on a sphere and conversely. We are now ready to extend this by studying general transformations of one plane to another. We will only examine linear transformations in detail. As will become obvious, these are convenient tools to describe the electrostatic potential field. Just as Euler and Lagrange were able to describe any rigid motion as a combined translation and rotation, so too, any linear transformation in the complex number plane can be described as a combination of translation, rotation, and, as we are not constrained by rigidity, stretching. The simplest would be to describe the w-plane as a function of the z-plane, where the complex number z has a real or x value and an imaginary y value: $z=x+iy$ and similarly $w=u+iv$. On the w-plane, the u-axis and v-axis, correspond to the x and y axes, respectively.

We can choose, as an example, the operations $w=z+b$ and $w=az$. The first will be a translation, the second a rotation. If we let both a and b equal, let's say, $3+2i$ for purposes of investigation, we can then compare the behavior of two parallel lines, $y=x+2$ and $y=x$, under these transformations. In the first transformation, a translation $w=z+3+2i$, the lines will simply be shifted by the addition of the vector (3,2); in the second $w=(3+2i)z$, they will be rotated. In both instances, the lines will remain parallel. They will intersect at the ideal point at infinity. [Figure 24]

Now we can use an inversion of measure relations as we mentioned

Figure 24

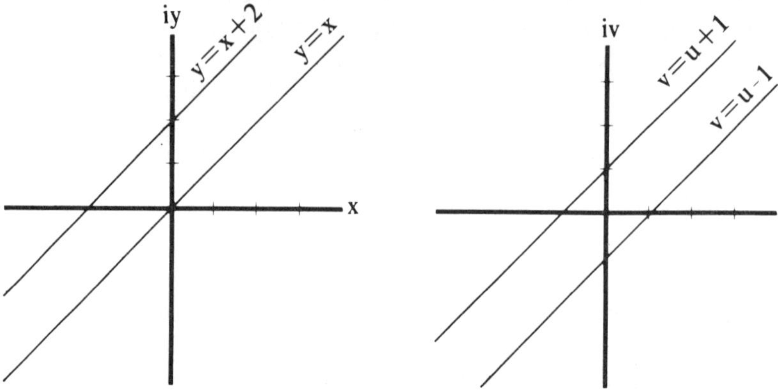

A and B. A simple transformation from the z=x+iy plane (A) to the w=u+iv plane (B), using the transformation w=z+3+2i. The line y=x maps into the line v=u-1, while y=x+2 maps into v=u+1. The only fixed point in this mapping is the point at infinity.

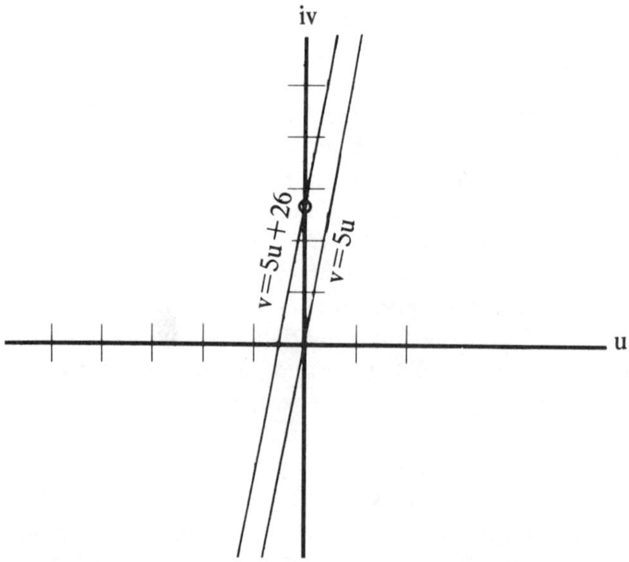

C. Using the transformation w=(3+2i)z, the line y=x maps into v=5u, while y=x+2 maps into v=5u+26. In this case, the points at both zero and infinity are fixed.

earlier: $w=1/z$. It is useful to refer this transformation directly to polar coordinates. We see that if $z=r(\cos\Theta+i\sin\Theta)$, then w will be transformed to

$$w=\frac{1}{r}[\cos(-\Theta)+i\sin(-\Theta)] = \frac{1}{r}(\cos\Theta - i\sin\Theta).$$

The sine of a negative angle is the negative of the sine; the cosines are equal.

The transformation can be made in two steps. Step one requires that we take the ray or vector z and reduce it to the reciprocal of its length $1/r$. In step two, we take the ray and invert it so that it now makes the angle Θ with the x-axis. We take the conjugate value of the ray which we obtained in step one. [Figure 25] The mapping $1/0$ to ∞, and $1/\infty$ to 0 follows from this as is shown most clearly when we refer the transformation to the sphere.

Up to now, we have looked at the mapping of the z-plane to the w-plane as point-to-pont or line-to-line transformations. Now we wish to look at the transformation of space as a whole. A transformation can have at most two fixed points, otherwise it leaves both planes identical. A transformation will have at least one fixed point, although this may be the point at infinity as in the case of translations. The mapping, $w=az+b$, will have two fixed points: the point at infinity and the point

$$m=\frac{b}{1-a}.$$

Such a mapping may be treated as a three part operation which brings m to the origin, performs a rotation and a stretching, and, then, translates the origin back to m. This makes it clear that a pencil of straight lines (lines which radiate from a point) through m goes over into itself as a whole, as does a family of concentric circles with m as their center. [Figure 26] We can, if we wish, choose any two points as fixed points and perform a mapping of the form

$$\frac{w-m_1}{w-m_2}=A\frac{z-m_1}{z-m_2},$$

where $A=r(at)(\cos bt+i\sin bt)$ and t varies between 0 and ∞. The circles passing through m_1 and m_2 are the equivalent of the pencil of rays passing through the fixed point in the first example. As we allow A to vary while keeping b fixed at zero, a point p on one of these circles will move between the points m_1 and m_2. A field will be set up in which these fixed points act as centers of attraction and repulsion of varying strength depending upon the function A. When we give b a value other than zero, we have introduced a rotary motion around the respective centers. [Figure

Figure 25

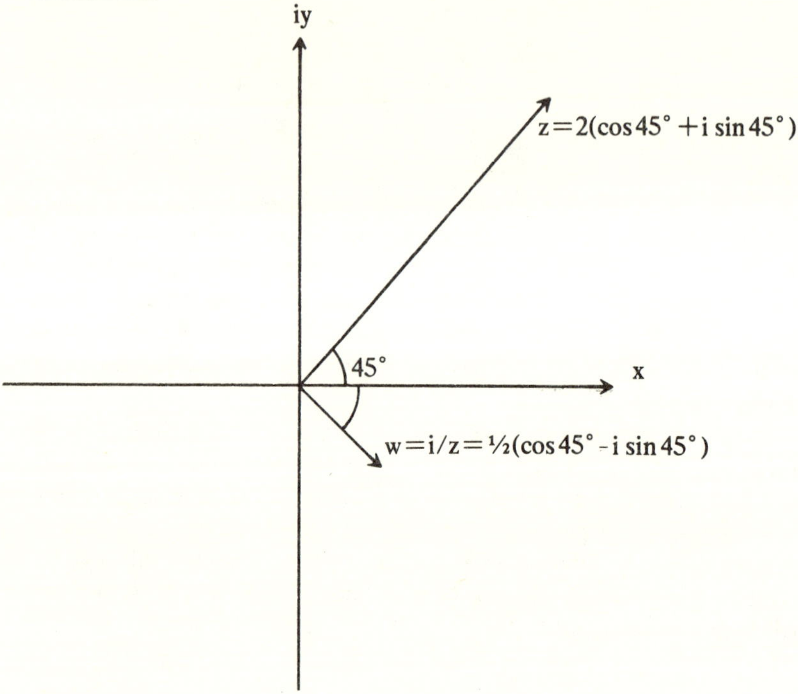

Inverting a complex number of the form z=A(cos Θ+i sin Θ).

$$\frac{1}{z}=\frac{1}{A\,(\cos\theta+i\sin\theta)}=\frac{1}{A}\left(\frac{\cos\theta-i\sin\theta}{\cos\theta-i\sin\theta}\right)\frac{1}{\cos\theta+i\sin\theta}$$

$$=\frac{1}{A}(\cos\theta-i\sin\theta)\,.$$

The magnitude of the inverse number is the inverse of the magnitude of z, while the new direction corresponds to a clockwise rotation through an angle 2Θ.

26] In this way we have the possibility of treating variations in the path of a charged particle through a magnetic and an electrostatic field by studying them as complex functions. The circulation can be used to indicate the path of an electron around the axis of a magnetic "field line." The condition that the force is perpendicular to the particle's velocity acts as a radial pull. In order to go beyond the electrostatic field, we will briefly look at electromagnetic potential.

The Lorentz force is an abstraction originally from the behavior of

Figure 26

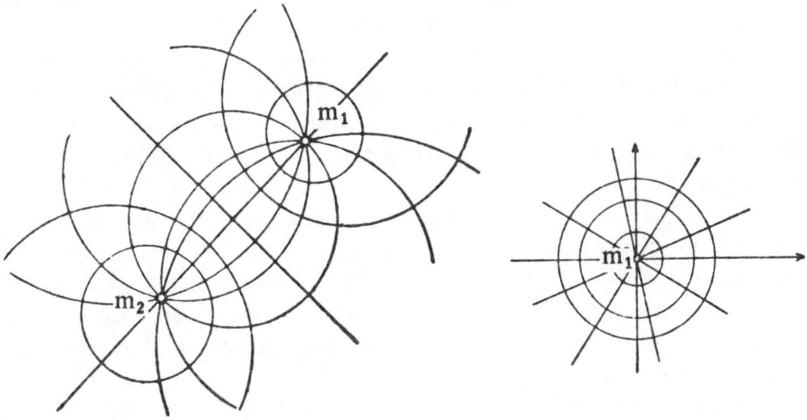

The transformation,

$$\frac{w - m_1}{w - m_2} = A\frac{z - m_1}{z - m_2},$$

which carries the point m_2 to infinity, stretches the circles passing through m_1 and m_2 into radii originating at m_1. Circles about m_1 become concentric. There is a physical parallel of streamlines (radii) and equipotentials (circles) to the motion of a point along the radii.

currents of electrons flowing through conductors, although it was subsequently proven that electron beams traveling through a dielectric are also subject to the Lorentz force, with appropriate relativistic corrections as their speed approaches the velocity of light. Riemann had an electron theory, although experimental technique in his day limited him to observing the behavior of currents running through wires. We will follow him in this experimental technique, which should not impose too great a practical handicap in fashioning experiments to demonstrate the magnetic effects generated by flowing electrons.

Cauchy-Riemann Equations

After Oersted's discovery that a magnetic needle laid over a wire will be deflected to a transverse direction by a current passing through the wire, it was but a short step further for Faraday to show that a compass taken around the wire will rotate its position. The original position, which the needle assumes, is directly connected with the direction of the current. If the current flows from south to north then the needle will align itself so that it points to the east. The rotation will be in a counterclockwise direction. [Figure 27] In these cases, we are assuming that compensation is made in the experimental design for the influence of the magnetic north of the earth. At any given position in space near to a wire through which current is flowing, a magnet will receive an impulsion to align itself so that, if it is taken in a complete circuit around the wire, it will complete a 360 degree rotation.

Faraday described this rotational tendency manifest by the magnet in the field of itself and the wire as a circular line of force in contrast to linear gravitational and electrostatic forces. The analogy is inappropriate, because no potential gradient sends the magnet "rolling" around the wire. Its tendency is to align itself with the magnetic field line (or, but this is not in question here, to be attracted to the wire). Faraday, then, reified his "circular line of force." This allowed him to visualize the circular lines of force around the wire apart from the interaction of the magnet and the wire, and, finally, to see them as an inherent property of a preexisting field. In this view, these preexisting lines of force merely encircled wire when current flowed. As Figure 28 shows, the actual gradient describing the magnet's rotational alignment is a radial pencil of lines.

With this in mind, we are ready to look at the Cauchy-Riemann equations. We have discussed the stereographic projection, which preserves directionality. We have looked at linear transformations which are also isogonal (angle preserving). We now ask the question: What is true of the general class of transformations which are isogonal and which do not distort proportionality over small areas? These are known as conformal mappings and they are characterized by the fact that they possess a derivative. A function which is differentiable in a region is called a regular function; the region is known as a regular region. A region may be regular except for one singular point — the case of the field and the charged particle. The Cauchy-Riemann equations give general conditions for determining whether or not a region is regular. We will first give a simple illustrative example and then discuss their importance. Let $w=z^2$, then

$$\frac{dw}{dz}=2z.$$

Express z as $x+iy$ and w as $u+iv$, then $(x+iy)^2=(x^2\text{-}y^2)+i(2xy)$.

$$u=x^2\text{-}y^2, v=2xy$$

$$\frac{\partial u}{\partial x}=2x=\frac{\partial v}{\partial y}, \quad \frac{\partial u}{\partial y}=-2y=-\frac{\partial v}{\partial x};$$

Figure 27

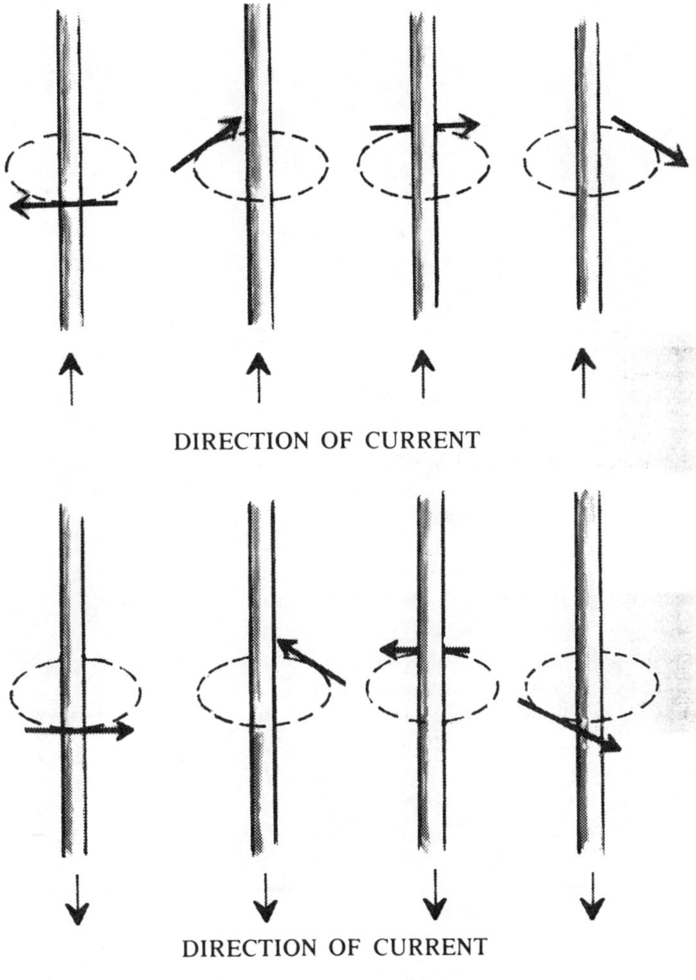

A magnetized needle orients itself in the direction perpendicular to the current flowing in a wire.

$$\frac{\partial^2 u}{\partial x^2}+\frac{\partial^2 u}{\partial y^2}=\frac{\partial^2 v}{\partial x^2}+\frac{\partial^2 v}{\partial y^2}=0.$$

The conditions are that

$$1)\ \frac{\partial u}{\partial x}=\frac{\partial v}{\partial y},\ \frac{\partial u}{\partial y}=-\frac{\partial v}{\partial x}$$

Figure 28

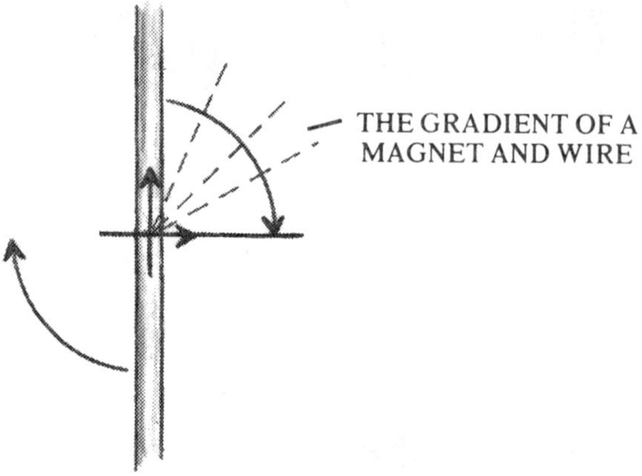

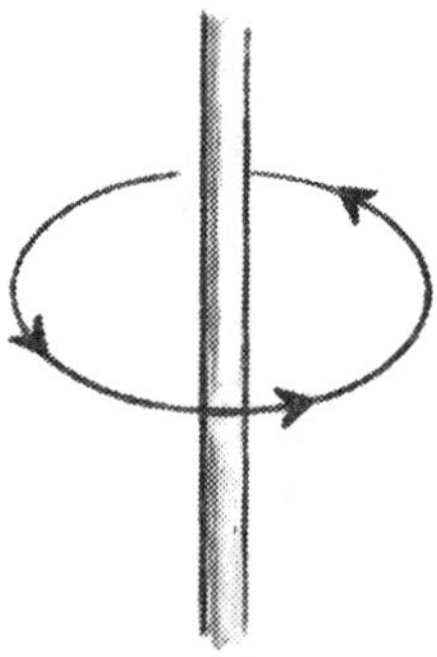

Equipotential lines of a magnet free to rotate are radial as shown by the dotted lines and should not be confused with the magnetic field lines shown circling the wire. This can be seen if you hold a compass against a wire which has a current running through it. In this case, the needle within the stationary compass will align itself in a direction which is perpendicular to the direction of current flow.

---

and

$$2)\quad \frac{\partial^2 u}{\partial x^2} + \frac{\partial^2 u}{\partial y^2} = 0, \quad \frac{\partial^2 v}{\partial x^2} + \frac{\partial^2 v}{\partial y^2} = 0.$$

One has only to consider that u represents a potential field and we see that the Cauchy-Riemann equations describe the conditions under which an integral can be taken between two points without regard to the path which is chosen. This is the condition which guarantees equivalently zero space curvature, rendering length measurement independent of path and conservation of energy. The condition that the Laplacian is zero defines that portion of the field in which there are no singularities. Physically, a singularity corresponds to the presence of a source of potential energy, such as an electric charge, at the point where the singularity is located. Having defined a regular region, the task which Riemann then set himself was to determine how to integrate singularities into the domain of mathematics.

Riemann's general approach to connecting complex functions with potential theory is described by Felix Klein in the following quotation from an 1894 lecture, "Riemann and His Significance for the Development of Modern Mathematics":

> Riemann begins with certain differential equations which are satisfied by the functions of x + iy. This takes on an immediately physical form. Let us put f(x + iy) = u + iv. Then, because of the indicated differential equations, the individual constituents u as well as v can be considered as a *potential* in the space of the two variables x and y. Briefly, one can characterize Riemann's discoveries by saying that he applies to these individual constituents the fundamental theorems of potential theory . . . He gave potential theory a fundamental significance for the entirety of mathematics and furthermore arrived at a series of geometrical constructions or, as I would rather say, geometrical inventions . . . A first step is that Riemann throughout regards the equation u + iv = f(x + iy) as a mapping of the xy-plane unto a uv-plane. This mapping turns out to be conformal, ie., angle preserving, and can in fact be characterized by precisely this property. We thus have a new auxilliary means for the definition of functions x + iy. In this context, Riemann develops the brilliant theorem that for an arbitrary simply connected region of the xy-plane and an arbitrary connected region of the uv-plane there always exists a function f which maps the one unto the other. This function is completely determined up to three arbitrary constants.
>
> In addition to that, he lays the foundations for the notion of a Riemann surface (as we call it today), that is, of a surface several sheets of which are extended above the plane and whose sheets are connected at so-called branch points. This without a doubt was the most difficult, but also the most successful step... The Riemann surface provides the means for representing the course of multivalued functions. On it, there exists the same kind of potentials as in the ordinary plane and their lawful behavior can be researched by means of the same tools. Furthermore, the method of conformal mappings remains in effect.

Thus, to repeat, in the simplest case u can be considered as an electrostatic potential. This can be represented as a family of concentric circles which surround the point of infinity which has become the center.

Figure 29

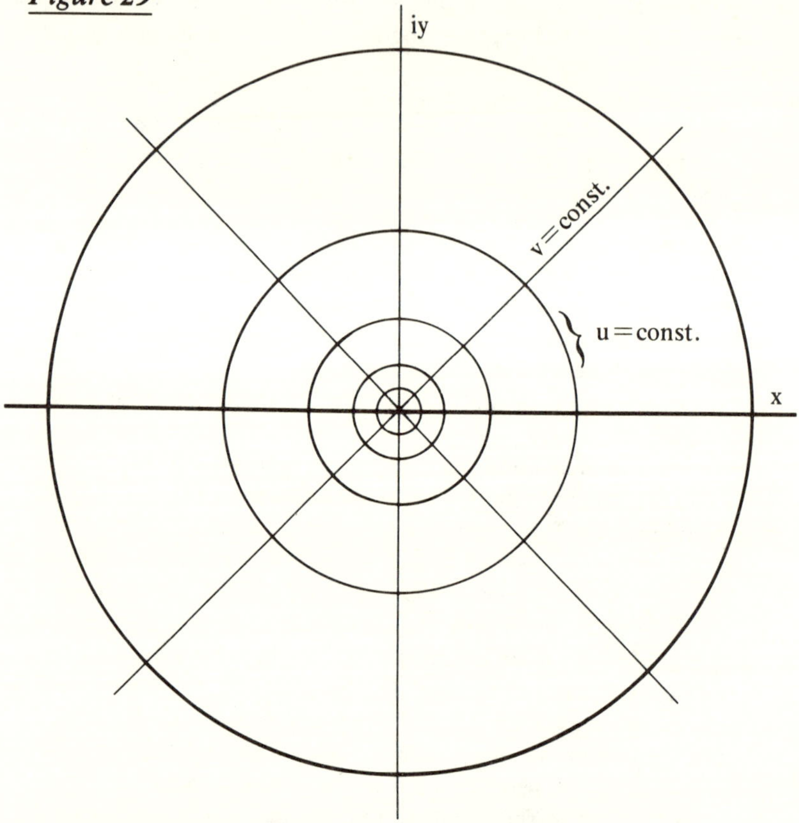

The application of complex functions to potential theory. Let a function w=u+iv, z=x+iy, w=f(z), then, if the Cauchy-Riemann equations hold, u can be interpreted as a potential function, with u=constant being the equipotential curves, and v=constant being the streamlines.

These are the curves u=constant. From this point of view, the curves v=constant, which are orthogonal trajectories, are streamlines. [Figure 29]

An even more interesting application applies when we consider the implications of the magnetic effect of current. Two wire loops through which current flows or one such wire and a magnet will affect each other as two magnets. This magnetic attractive force is known as the ponderomotive (or magnetomotive) force. This force will be directly proportional to the strength of the currents. In this case, we can consider the lines as magnetic potential, and the circles, u=constant, can then be considered

Figure 30

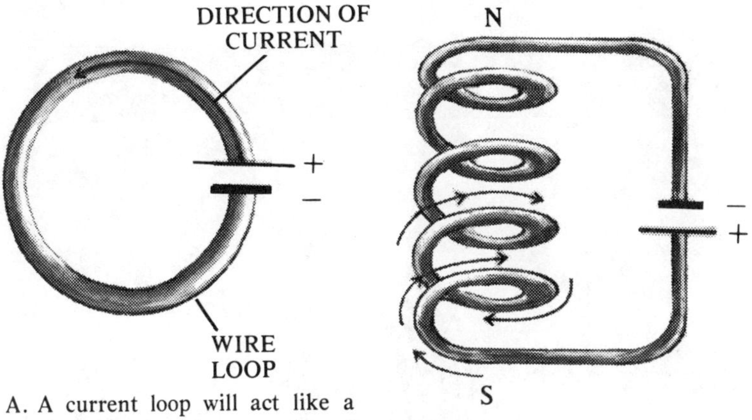

A. A current loop will act like a magnet whose poles are aligned perpendicular to the place of the loop and which resides at the center of the loop.

B. Multiple loops (a solenoid) multiply the strength of the magnetic effect.

as streamlines. The loops will be drawn toward or away from each other, depending upon whether the current runs through them in parallel or antiparalled directions. [Figures 30, 31, 32] It is to Felix Klein that we owe the direct interpretation of the Cauchy-Riemann equations in terms of potential theory. He also suggested another "projective" transformation which describes the electromagnetic field and currents as interconnected screws and wrenches. We will discuss this transformation in the next chapter. The point to be born in mind by the nonmathematical reader is the essentially geometric nature of potential field theory.

The Cauchy-Riemann equations define how a region with one or more singularities can be integrated. Essentially, the method is simply to treat these singularities as holes which can be eliminated by suitably chosen paths. [Figure 33] The method is to cut the region in such a way as to prevent the complete encirclement of any one singularity. It is something like turning a donut into a frankfurter and then slicing it almost, but not quite through to make sure that there is only one way to encircle it end-to-end. [Figure 34]

Riemann's most important single invention was his method of connecting a sequence of ordered manifolds through branch points. In its simplest application, this is a method which allows us to express a function such as $w=\sqrt{z}$ by a continuous path through to separate, but connected z-planes. A singularity exists at the point $z=0$. The idea is

Figure 31

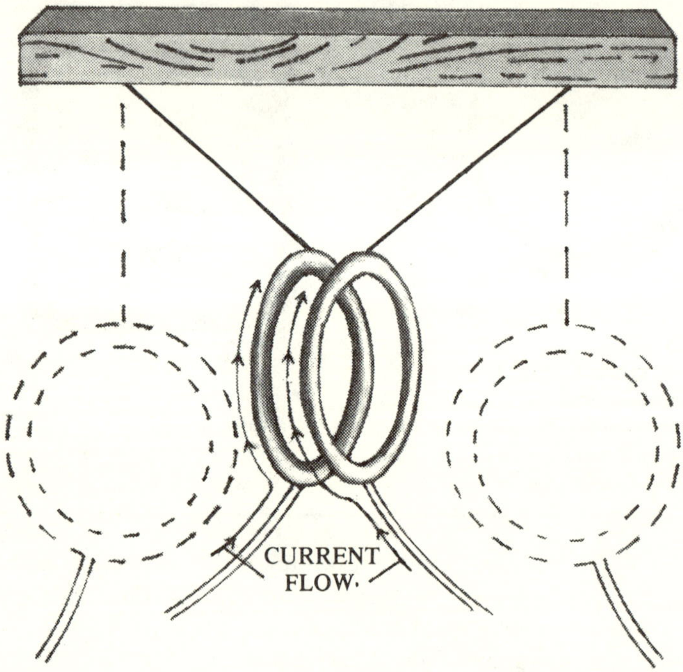

Two current loops will be attracted to each other if the currents are aligned parallel; they will repel if the currents are antiparallel.

understood easily enough by extension from the example of a parabola in the Euclidean plane. Suppose that the parabola has the x-axis as its axis. For every given value of x, such as nine, there are two possible y values: positive and negative three. Riemann's extension of this idea to the complex plane involves an interlocking of singularities to represent a complex situation of actual determining choice. Each possible manifold defines a future series of unfolding manifolds. A set of such manifolds may collectively represent mere variation on a given theme, alternate proposals to achieve broadly the same end by slightly differing means; or the manifolds may represent totally different geometries: survival or destruction. The path through such singularities must subsume the concrete intersection of the geometry of now and then-to-be.

Figure 32

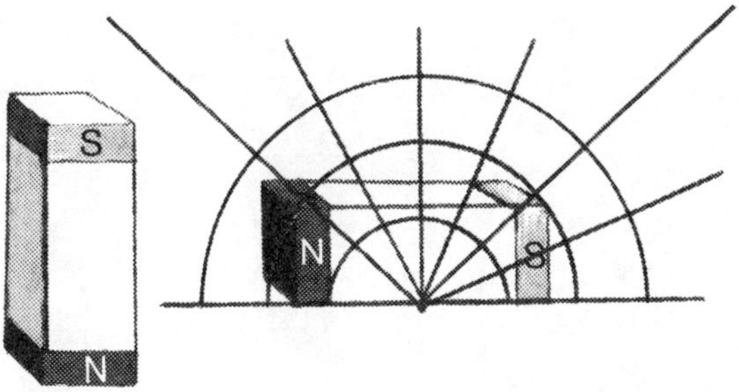

A. A magnet free to rotate in the presence of another magnet will align itself in a minimum energy configuration. The magnet on the left is fixed in both cases. Equipotential lines radiate from a common center, while streamlines appear as circles perpendicular to them.

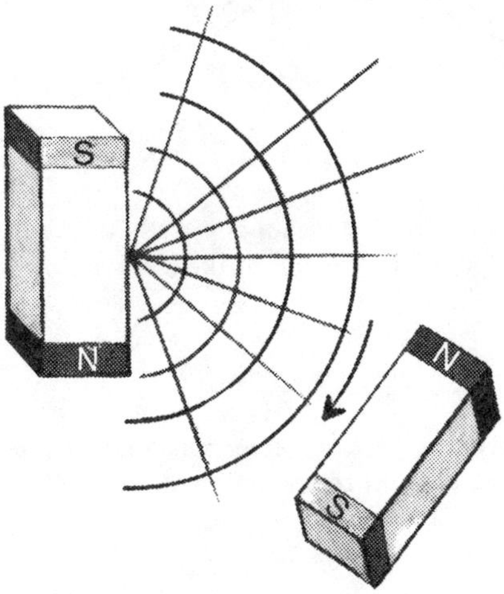

B. When moved sideways, the magnet will continue to make incremental rotations to maintain the minimum interaction energy.

The geometry of any given potential field is only knowable in the final analysis by reference to the higher-order potential fields which are implicit in its existence. But scientific practice is not mere universal bookkeeping. The choice as to which implicit geometry is realized, which potential branch is actualized, lies in the final analysis with us, the willful scientists.

The foregoing geometric treatment, and, in fact, any particularization of so-called Riemannian geometry, does not and cannot describe a

Figure 33

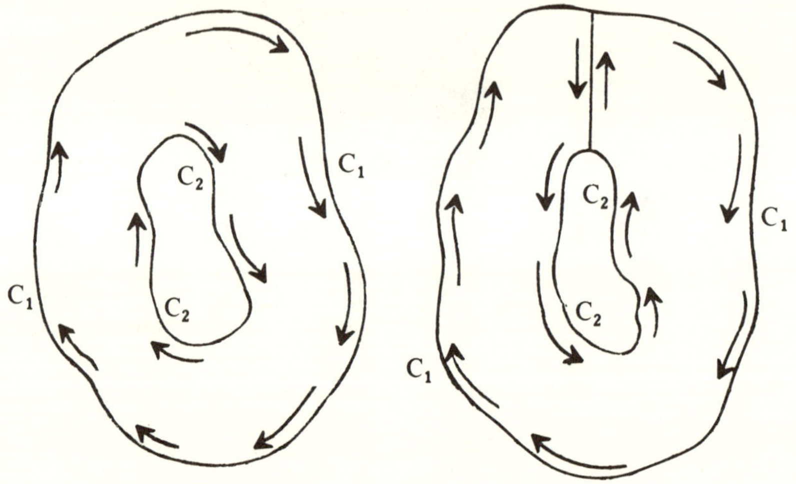

If a function f(z) is analytic inside the region bounded by the curves C_1 and C_2, the integrals along C_1 and C_2 are equal, if the direction of the integration is the same,

$$\oint_{C_1} f(z)\,dz = \oint_{C_2} f(z)\,dz.$$

By making a cut which connects the two curves, the analytic region can be traversed by a single integral, with the direction of the C_2 part changed,

$$\oint_{-C_2} f(z)\,dz = -\oint_{C_2} f(z)\,dz$$

or

$$\oint_{C_1} f(z)\,dz + \oint_{-C_2} f(z)\,dz = 0,$$

since the integrals along the cut cancel each other. Cauchy's theorem, that $\oint f(z)\,dz = 0$ in a region where f(z) is analytic, and this cutting process prove the equivalence of the integrals along C_1 and C_2. The curve C_2 can enclose a singularity in the function f(z), so that the theorem provides a way of evaluating the integral despite the singularity. This method, when extended, becomes the well known Residue Theorem.

Figure 34

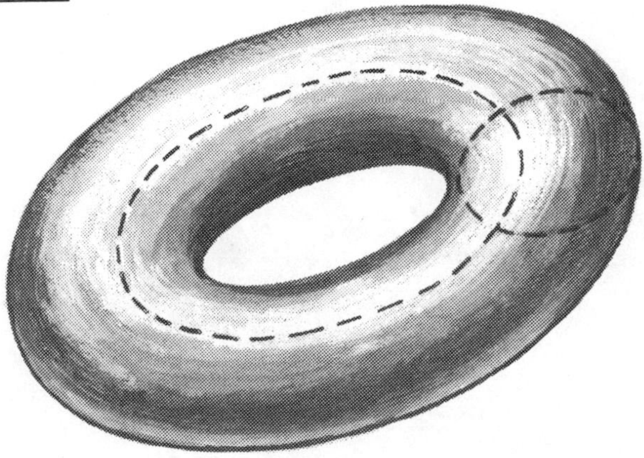

A donut or inner tube can be cut in such a way as to give it a topologically equivalent appearance to the region in which f(z) is analytic in Figure 33. First, by slitting it the long way, the boundaries C_1 and C_2 are created. Then, a slice across provides the cut which removes the encirclement of the singularity — the center of the donut. (Note: one can achieve the same effect by reversing the procedure.)

unified field, as Riemann himself knew well. Riemann's major accomplishment was to comprehend critically the limitations of existing potential theory by taking it to its furthest geometrical extremes. Nonetheless, Riemann's geometric treatment of potential theory allows us the most adequate approximation to a unified field theory based upon the experimental method of nineteenth century physics. (For a full presentation of Riemann's mathematical work, which lies outside the scope of this book, see Parpart, Uwe. "The Concept of the Transfinite." *The Campaigner.*)

CHAPTER V
Maxwell's Equations

Especially since the development of quantum electrodynamics, knowledgeable scientists have recognized the epistemological blunders and shortcomings of Maxwell's attempted theoretical unification of electromagnetic and light radiation. Nevertheless, the convenient justification is offered that only Maxwell's equations provided the mathematical-physical basis needed at the time for technological advances in wave transmission. This is not true. In fact, both the form of Maxwell's equations and their underlying epistemology are inferior to Riemann's earlier systematic elaboration of electromagnetism according to potential theory.

That Riemann's work was significantly more advanced than Maxwell's is verified by the explicitly relativistic form of Riemann's equations. As we shall show, Riemann brilliantly anticipated the Lorentz transformations, although not in detail, as his work preceded the Michelson-Morley experiments. Maxwell's equations, on the other hand, were capable of relativistic treatment only after being "translated" into the form of an electron theory by Lorentz.

Riemann's treatment was superior to Maxwell's precisely because he treated electrodynamics and electrostatics from the standpoint of a universalizing geometry in which electrostatics is subsumed under relative velocity — the case in which the relative velocity of two objects (either moving at the same speed or at rest) is zero. The demand for such coherency follows from his epistemology just as Maxwell's (and contemporary texts') willingness merely to build electrodynamics on the existing foundations of electrostatics follows from his reductionism.

Quantum electrodynamics merely extends this reductionist indifferentism to incoherence, asserting the necessary dichotomy between the wave and particle natures of the field. Just as Riemann utilized the experimental evidence of electrodynamics to reconceptualize the electromagnetic field, the task before us today is to go beyond Riemann and the

Relativity Theory which he foreshadowed by subsuming the electromagnetic field within the phenomena of radiation.

Maxwell's equations for the propagation of electromagnetic waves have the further flaw of failing to comprehend the source in their descriptions of radiation. Since Einstein's work rests upon these equations, his search for a unified field was on this account alone foredoomed. Furthermore, Maxwell dichotomizes between the magnetic and the electric force components of the electromagnetic wave so that these "forces" are treated by separate wave equations as discrete phenomena which are mysteriously interconnected.

There are two basic equations which express the fact that a magnetic field is produced by current flowing through a wire loop and that electricity can be generated in a loop by varying the magnetic flux through the loop. By extension of the first equation and the simultaneous algebraic solution of them, the equation describing the propagation of electromagnetic waves is derived.

We will develop Maxwell's equations from the context out of which they emerged historically.

Maxwell's theory, in the last analysis, rests on the work of potential theory as it was discovered and elaborated by Gauss and Weber, Ampère, Biot and Savart, and, lastly, Riemann. Each of these men treated the collectivity of magnet and current loop, or current loop and current loop either implicitly or explicitly in terms of the same notion of energy potential which guided Gauss's discovery of the electrostatic potential function.

In the case of the electromagnetic field, such a unique function of position does not exist. Instead, it is necessary to conceptualize a higher-order notion of energy potential which depends upon the velocity of charge (i.e., the current) as well as its position. This is known, in modern terminology, as the vector potential.

Ponderomotive Force

It took a full ten years of experimental work to discover the method of generating electricity after the continental physicists had discovered the (magnetic-attractive) ponderomotive force. Two wire loops, which are oriented parallel to each other, will be attracted toward or repelled from each other as magnets, as the electric current runs (respectively) in the same or opposite direction through them. After Oersted discovered that a magnet is affected by the field surrounding a wire through which current flows, it was but another step for Ampère, Biot, and Savart to discover that a coil of wire can form an electromagnet whose strength will vary proportionally as the strength of the current and as the tightness and number of windings in the coil.

There are several equivalent forms of the law which measures the force between two such wires, or a wire and a magnet. Ampère arrived at the empirical test of his calculations by arranging two coils with current flowing through them next to each other in perfect balance. He then was able to test their deviation when "disturbances" in the form of other coils were in their vicinity, and, in this way, to measure the magnetic effect of the coils which were introduced. Metal magnets and electromagnets can be equivalently calibrated in this way.

Biot and Savart announced the law which bears their name at a meeting of the French Academy of Science in 1820. Their law states that the action experienced by the north or south pole of a magnet placed at any distance from a straight wire carrying a voltaic current may, in their words, be expressed as follows: "Draw from the pole a perpendicular to the wire; the force on the pole is at right angles to this line and to the wire, and its intensity is proportional to the reciprocal of the distance." [Figure 35]

Exactly a week after Oersted's discovery, André-Marie Ampère showed that two parallel wires carrying currents attract each other if the currents are in the same direction, and repel each other if the currents are in opposite directions. To explain the action of the currents, he arrived at a form of the Lorentz force law: The ponderomotive force between two such currents is given by

$$\frac{ii'}{c^2}\int \mathbf{ds}'\times\frac{(\mathbf{ds}\times\mathbf{r})}{r^3},$$

where **ds** and **ds**′ indicate elements of the two currents and **r** is the vector distance between the current elements. This force is equivalently the vector product of the current element of one wire with the magnetic intensity of the other.

Reference to Figure 32 in the last chapter will show that the magnetic intensity **B** is equivalent in the transformation in which the streamlines are circular and the lines of equipotential are radial to the direction of the streamlines. The ponderomotive force is equivalent in the conjugate transformation to the radial streamlines. The magnetic force is a vector d**B** in the direction of the streamline.

A circuit of a test particle over a path which includes a singularity, in this case the current, will not be conservative. This observation becomes clear by examining the path dependence of the magnetic vector potential. [Figure 36]

The situation is similar to rolling downhill all the way and, yet,

Figure 35

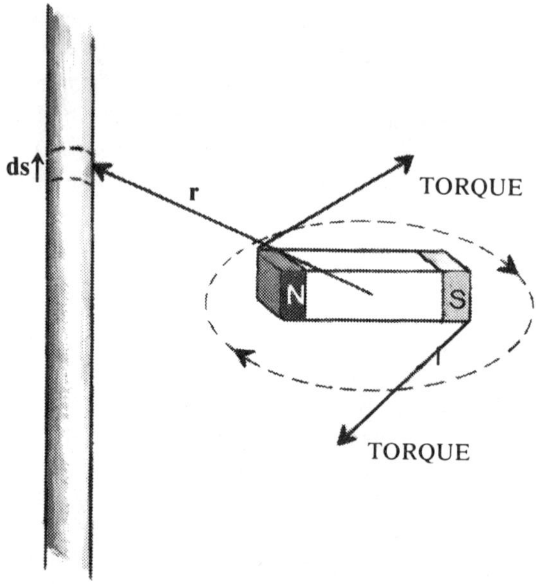

The Law of Biot and Savart indicates the strength of the force needed to align and attract a magnetic needle. The force acts like a magnetic field which is created around a current-carrying wire. In modern terms, the force is proportional to $i(\mathbf{ds} \times \mathbf{r})/r^3$, where i is the current strength and **r** is the distance from the magnet to the infinitesimal current element **ds**. The product $\mathbf{ds} \times \mathbf{r}$ — called a vector product — has a magnitude proportional to the product of magnitudes **ds**, **r**, and the sine of the angle between these two vectors.

Note to the experimenter: To carry out the experiment depicted here, the magnet on the right should be pivoted and, therefore, free to move. In this case, when current is run through the wire, the magnet will align itself in a direction perpendicular to the direction of the current, as shown in the picture above. If, however, the magnet is suspended on a string, when the wire is connected to the battery, the magnet will not only align itself, but will also be attracted to the current-carrying wire.

---

reaching the top, or climbing continuously upward and finding oneself back down at the bottom. This is seen when a magnet is carried around a current and rotates through 360 degrees as it constantly realigns its position. Energy is continually supplied to the field by the current. When the current is steady, this amount is represented by $\int \mathbf{B} \cdot \mathbf{ds} = 4\pi i$. This energy will flow out from the wire as we indicated in discussing the general flow of field energy in connection with Riemann's notion of the retarded potential which travels outward with a finite velocity — at the speed of light. The energy, of course, will ultimately be provided by some source equivalent to a battery. In contrast to an electrostatic potential field, which can be maintained at relatively constant potential for a considerable period of time, a battery runs down and must be recharged or, in other words, repotentiated. The relationship between the current and the potential of the field to perform work in a complete circuit around a wire is also represented by the formula **J=curl B**.[Figure 37] **J** represents the current. Curl is a word coined by Hamilton to denote the fluid rotation over a surface area. [Figure 38] Both Stokes and Riemann established that the superposition of rotations at every point in an area can be subsumed by the integral of the rotation around the perimeter.

Until now, we have only looked at the gradient of the field as it changed with location. The existence of a current, in a certain sense, introduces the notion of time, since current is charge in motion; nonetheless, in discussing the ponderomotive force and the magnetic intensity, we were still concerned with a gradient dependent upon position.

Through the formula for the mutual potential energy of two currents

$$\frac{ii'}{c^2}\int \frac{\mathbf{ds}\cdot\mathbf{ds}'}{r}.$$

we can derive the magnetic vector potential, which brings time dependence into the potential. The expression above indicates the potential for two currents to exert a ponderomotive force on each other. It is derived from the formula for the field of magnetic intensity of a current i flowing in a circuit **s** in this field,

$$\frac{i}{c}\int \mathbf{B}\cdot\mathbf{dS}$$

and the expression for **B** which can be written as

$$\frac{i'}{c}\int \mathbf{curl}\frac{\mathbf{ds}'}{r}.$$

We transform the integral

$$\frac{i}{c}\int \frac{\mathbf{ds}'\times\mathbf{r}}{r^3}$$

which gives the integral

$$\frac{ii'}{c^2}\int\left(\mathbf{curl}\frac{\mathbf{ds}'}{r}\cdot\mathbf{dS}\right)$$

Figure 36

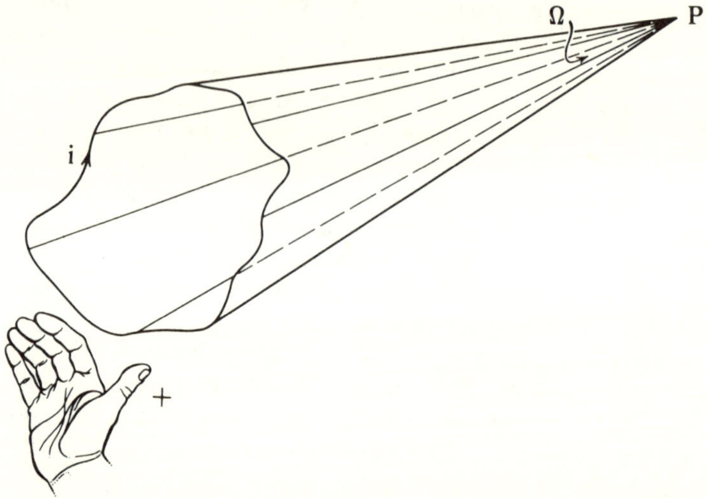

A. Ampère's Law for a closed loop can be expressed in terms of the gradient of the solid angle Ω subtended at the point where the magnetic induction is to be determined. As will become clear in what follows, the magnetic potential is determined only up to a constant which is an integral multiple of 4π.

---

which by Stokes's theorem can be written in the form with which we started the discussion. The quantity

$$\frac{i'}{c}\int\frac{\mathbf{ds}'}{r}$$

known as the vector potential and customarily denoted by the letter **A**, is a function of the position of the element **dS** from which r is measured. [Figure 39]

Induction of Electricity

We shall now consider the capacity of the field to perform work as a function not of the change of position of a particle in it, but as a function of the change of the field itself in time. We can study this by keeping our test coil at rest and either allowing a nearby current to vary with time or varying the distance between the current and the test coil. In either case,

Figure 36 Cont.

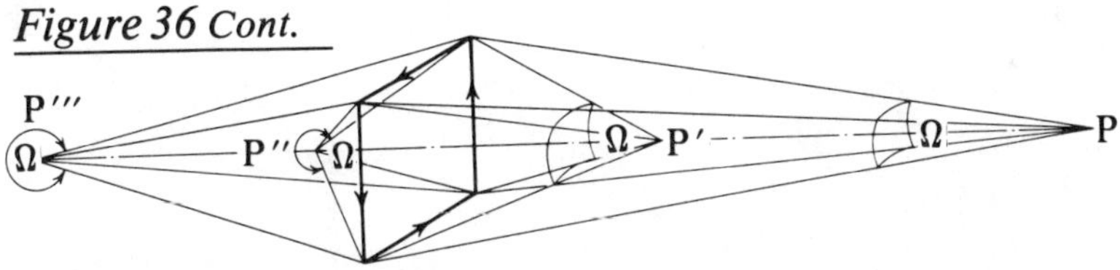

B. The point P, shown on the positive side of the loop, is allowed to migrate to P‴ on the negative side. If P starts from an infinite distance away to the right and P‴ is infinitely far away to the left, then Ω starts with a value zero, passes to 2π, and ends up at 4π.

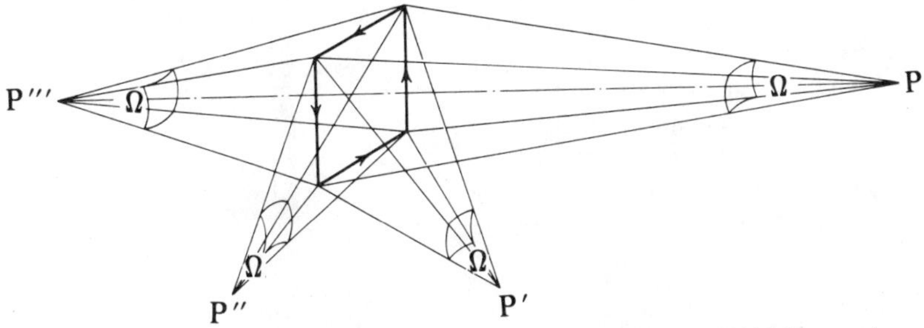

C. If instead of passing through the loop, P migrates to P‴ by passing around the outside, the final value of Ω will be zero. Consequently, we can say that Ω is only determined up to multiples of 4π, so that the potential which gives rise to the magnetic induction is only determined to the same degree.

we will produce work. Experiment shows that this work can be realized in the form of an electromotive force. Electric current can be produced in a test coil with a strength which will vary with the rate of variation of either the current or the distance, or both: an electromotive force will be produced as a function of the rate of variation of potential with time. We have thus restated the law of induction of electricity in terms of potential theory. An equivalent account can be given if electricity is produced by the variation of a magnet.

It is Faraday who first discovered this relationship; although the accidental character of the discovery is testimony to his unfortunate theoretical impoverishment. Faraday had been seeking to generate electricity for the ten years which preceded his discovery in 1831. He reasoned that a current should induce an oppositely directed current in a coil placed near it, by analogy to the induction of static electricity. He completely failed to comprehend the difference between energy and momentum. In order to keep a current circulating at ordinary temperatures, it is necessary to supply a continuing amount of energy to compensate for the resistance

Figure 37

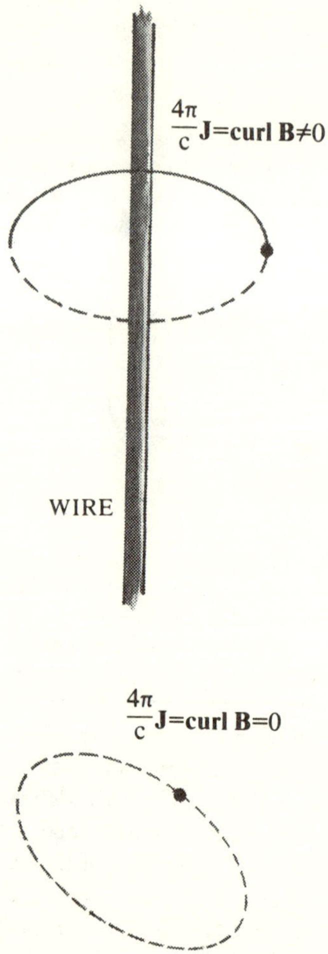

A. The vorticity of the magnetic field can be imagined as the tendency for a magnetic molecule, a hypothetically isolated monopole, to make a whole circle about a current-carrying wire as the magnet aligns itself to the wire. The strength of the circling tendency is proportional to the strength of the current in the wire.

B. In a region devoid of currents, the circling tendency also vanishes. By contrast, the electrostatic field always has vanishing vorticity. Referring to the discussion of Figure 36, any closed path for an electrostatic field leads to the same value of the potential function on return to the point of departure. This will only be the case for a magnetostatic field, if the path does not enclose a current.

within the conductor, manifest mainly as heat. Although energy is conserved in this situation, as in the case of induction of static electricity, momentum is lost through resistive friction, while, in the static case, the momentum is identically zero at all times.

Because he looked at matter as inert, he was systematically unable to understand the field-particle collectivity. Electrostatic induction is an

Figure 38

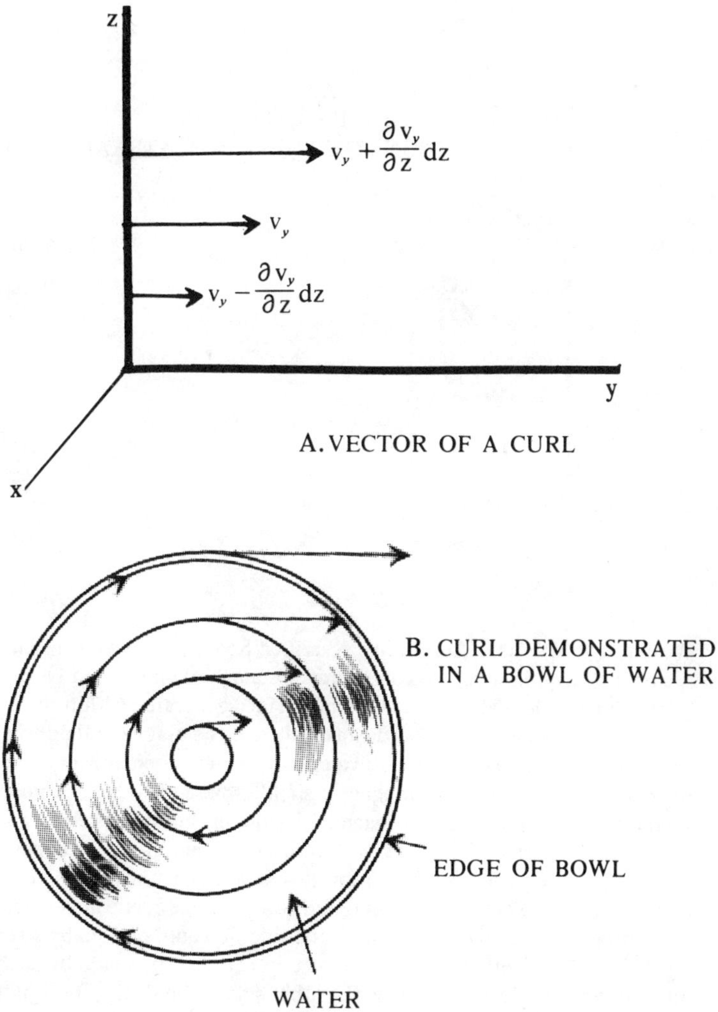

A field, such as a velocity (v) field, for which **curl v** $\neq 0$ is said to be a vortex field. The nonvanishing vorticity illustrated corresponds to a rotation about the x-axis (A), which could be, for example, the motion of water in a stirred bowl (B) or as it goes down a drain. The increase in values of v_y with displacement along the z-axis gives rise to a circulating tendency in the field.

Figure 39

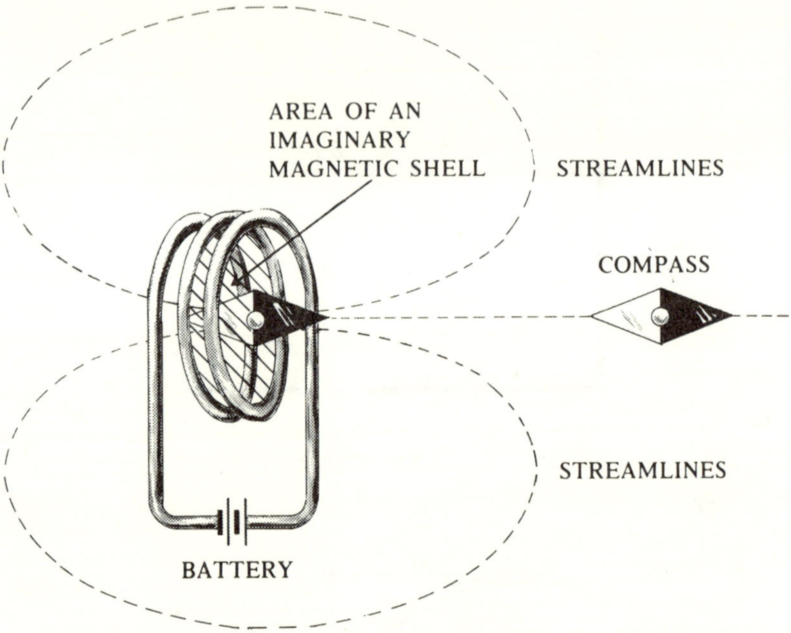

We can envisage the discontinuity described here by a simple experiment. A current carrying wire, which generates a magnetic field, can be said to have a "magnetic shell" defined by the surface generated within the plane of the wire. Where Faraday and Maxwell treat the magnetic field lines as closed figures, topologically equivalent to circles, Riemann emphasizes their discontinuity at the "magnetic shell" where they have a point of singularity. The lines may approach both sides of the shell as closely as we wish, but the surface itself has two different potential values depending upon which side we approach it from, just as a magnet has two poles. The point is clear if we envisage an imaginary unipolar magnetic molecule. If it was strongly attracted to the shell on one side, it would be equally strongly repelled from the other. A continuous trip can only be made by such an imaginary molecule if it travels to the other side of the shell — by a path or streamline which goes outside rather than through the shell. The question of the existence of photons where there is no charge makes this sort of distinction more than moot. Faraday claimed substantiation for his theory of magnetic poles on the basis of an experiment in which a magnetic needle floated through the inside of a wire electromagnet. In various experiments which we have made, we have noted the circumstance that the magnet stopped at the center of the solenoid, which is what we predict would occur.

impulse to alignment which is transmitted as a field travels outward from a source. Electrostatic induction represents the relatively fixed separation of charge into a relatively stable equilibrium position. While motion continues on the level of microscopic interchange, on the macroscopic level, the field can be considered to be at rest once induction has occurred. A current represents a dynamic equilibrium in which charge is constantly in motion.

The magnetic field is a function of this dynamic equilibrium, a new degree of freedom representing a higher-order energy potentiation. The generation of electricity as a function of the variation or motion of the magnetic field is again a higher order of potentiation. The dynamic equilibration of this variation — a pulsed variation of the magnetic field, an oscillating variation of current — which implies the alternating acceleration and deceleration of electrons, produces that still higher order of potentiation — electromagnetic radiation. Properly these orders of potentiation do not represent new geometries since, for example, the electrostatic field has a separate identity only by abstraction or initially through man's ignorance. Cantor's distinction between ordinal and cardinal numbers offers a useful heurism; the electrostatic and magnetic fields can be viewed as ordinal transfinite numbers subsumed under the cardinality of the electromagnetic field. (Within every transfinite number class above the first, cardinal numbers are no longer identical with ordinal numbers as is the case with finite numbers. Ordinal numbers can be considered as variations — Riemannian surfaces branching from a given singularity — all represented by a given function which represents an overriding geometry.)

Faraday correctly reasoned that since a current can affect a magnet, the reverse should also be true. He incorrectly looked for this effect as a simple function of position. This followed from his conception of streamlines as the dynamic agent of transmission of impulse, in effect as substance in motion. He conceived of streamlines as discontinuous circular tubes which resembled the pattern made by iron filings under the influence of a magnet. [Figure 40] He did not understand that streamlines merely represent the potential gradient of a surface — a capacity of the field to accomplish work. In the course of his investigations, Faraday discovered the actual law. It is by continuous variation of the electromagnetic potential field that current is generated.

Faraday discovered that he generated current in a nearby secondary coil when he started or stopped the current in his primary coil. This current in the secondary coil was of extremely short duration, at the instant of connection or disconnection of the first coil with a battery. It was also exceedingly weak. [Figures 41 and 42]

At around the same time, he also experimented with the then-

Figure 40

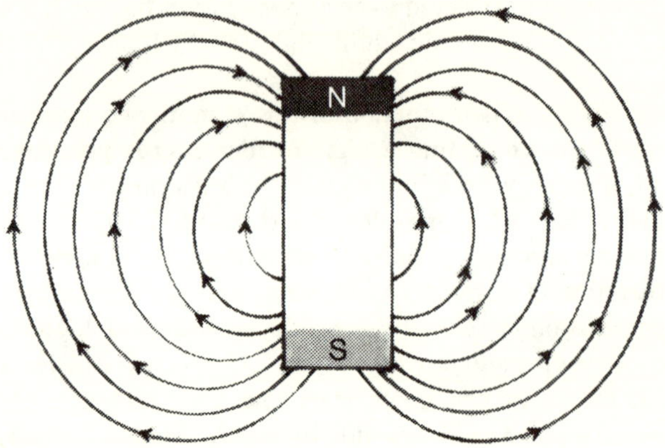

Any effect produced by the primary coil can be produced by a magnet, provided that the streamlines (lines of magnetic induction) are aligned in the same manner as those of the primary coil. The familiar pattern of iron filings spread thinly over a paper placed over a magnet is of interest. It was the concentration of the filings into lines like those drawn above which led Faraday to his reification of lines of force. It should be noted through repeated experiment that the lines produced are not identical each time, but are approximately similar. The filings create a situation in which the homogeneous potential gradient represented by the streamlines becomes discontinuous through the interaction between the filings and the field. This interaction creates new areas of concentrated field energy and others of correspondingly low energy density. This reordering of potential is reminiscent of the far more pronounced self-ordering of plasmas.

---

puzzling phenomenon of Arago's wheel. When a copper disk was rotated on its axis, a magnet placed over it also rotated. (Any other conductor can, of course, be used in place of copper.) Explanations were varied, with Biot even hazarding the guess that the copper became magnetized as a result of centrifugal separation of the magnetic fluid contained within it. What occurs in Arago's wheel is the generation of electrical-eddy currents in the disk, with the potentiation on the outer rim being lower than that at the center due to their different angular velocities. The general direction of current flow is radial, producing a situation in which the magnetic fields of the magnet and disk cannot be aligned. Faraday demonstrated the existence of the current. [Figure 43 and 44]

Figure 41

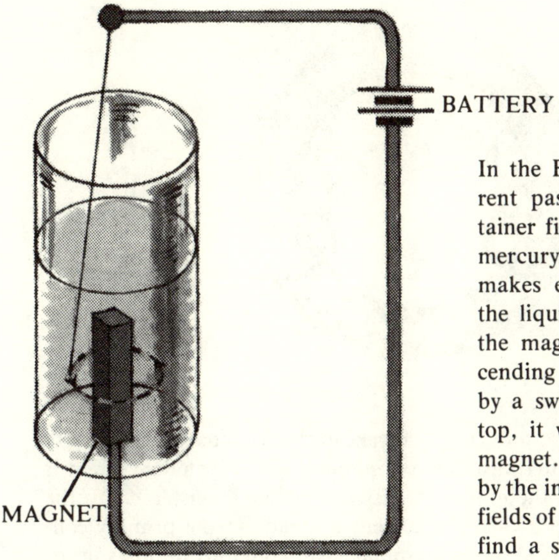

In the Faraday motor, a current passes through the container filled with salt water or mercury. One end of the wire makes electrical contact with the liquid near the bottom of the magnet. If the wire, descending into the cup, is fixed by a swivel connection at the top, it will rotate around the magnet. The rotation is caused by the inability of the magnetic fields of the magnet and wire to find a stable equilibrium.

He went further to construct an apparatus in which wires were affixed to the copper disk while a magnet rotated beneath it. This time nothing happened. While all of his connections were previously made through cups filled with mercury and were thus sliding rather than fixed connections, in this case, as the disk itself did not move, the connections remained fixed. When he reconnected the disk and magnet, and simultaneously rotated them, he again picked up a current. The conditions of this experiment were varied by the relationship of the test wire to the disk. However, he made the generalization from his results that the magnetic field remains fixed despite the rotation of a magnet. This only served to confirm his notion that magnetic field lines exist independently of sources to which they simply attach themselves.

Faraday correctly identified the law of electrical induction only to reify it. He wrote:

> Whether the wire moves directly or obliquely across the lines of force, in one direction or another, it sums up the amount of the forces represented by the lines it has crossed so that the quantity of electricity thrown into a current is directly as the number of curves intersected.

Figure 42

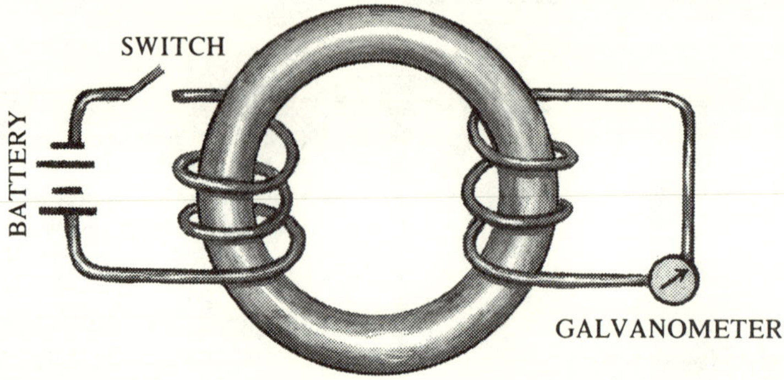

Faraday's apparatus for inducing a current in the secondary winding. He expected the iron ring to transmit a continuous effect. When the current is switched on or off, there is a transient current registered in the galvanometer. However, when the current is steady in the primary coil, there is no effect observed in the secondary. He noted his findings in his diary:

> Continued the contact of A side with battery, but broke and closed alternate contact of B side with flat helix. No effect at such times on the needle. Depends upon the change at battery side. Hence, is no permanent or peculiar state of wire B but effect due to a wave of electricity caused at moments of breaking and completing contacts at A side.
>
> Tried to perceive a spark with flat helix at juncture B side but could find none. Wave apparently very short and sudden.

He visualized this in terms of sweeping the lines of force into a basket — so many collected per unit of time. This, of course, merely reifies the appropriate notion of a time gradient of potential which is contained in Minkowsky's development of relativity theory and the notion of four-dimensional space.

Expressed in terms of the magnetic flux of either a magnet or a current, the law is quite simple. The flux through the surface bounded by a coil can be written as $\int \mathbf{B} \cdot \mathbf{N} da$. [Figure 45] The change in flux will be

$$\int \frac{\partial \mathbf{B}}{\partial t} \cdot \mathbf{N} da.$$

Figure 43

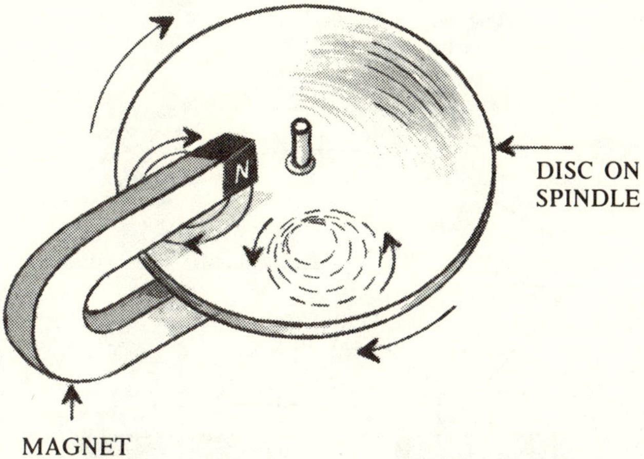

Arago's wheel demonstrates the existence of eddy currents. The wheel, a conducting plate, is rotated producing an induction effect. As the magnetic field strength varies in a region of space (a region of the wheel), an electric field is generated, which in a conductor gives rise to a current circulating in a section of the wheel. The current in turn interacts with the wheel to slow it down. If the interaction of the current and the magnet acted to accelerate the wheel's rotation, this would again be creation of energy *ex nihilo.* An electrical generator is the extension of this experiment. Contacts can be brought up to the wheel and the eddy current can be tapped. Alternately, the production of eddy currents in this device can be explained in terms of the Lorentz force law ($\mathbf{V} \times \mathbf{B}$). Electrons moving through the magnetic field feel a force which causes them to revolve in the plane of the plate.

This will equal the negative of the electromotive force since the current generated in the secondary coil will be in the direction opposite to that of the primary when the primary current is increased. The notation da represents surface area; **N** represents the direction normal to the surface. Stokes's law allows us to make the transformation

$$\int \mathbf{E} \cdot \mathbf{dl} = \int \mathbf{curl\ E} \cdot \mathbf{N}\mathrm{da} = -\frac{1}{c}\int \frac{\partial \mathbf{B}}{\partial t} \cdot \mathbf{N}\mathrm{da},$$

so that we may write

Figure 44

The galvanometer, electric motor, and electric generator all operate due to the interaction of magnets and currents. In fact, the galvanometer is basically an electric motor.

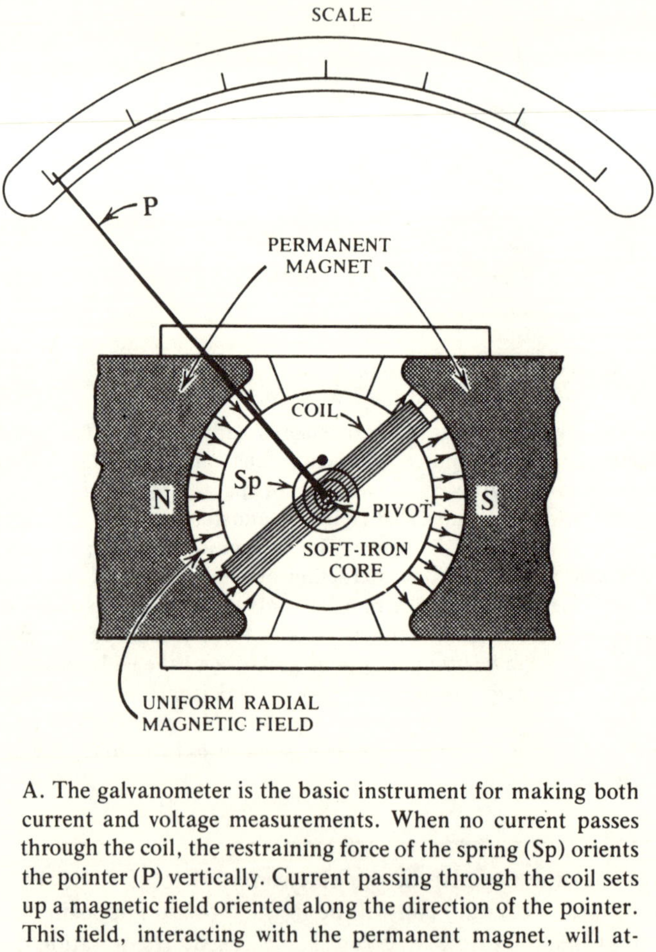

A. The galvanometer is the basic instrument for making both current and voltage measurements. When no current passes through the coil, the restraining force of the spring (Sp) orients the pointer (P) vertically. Current passing through the coil sets up a magnetic field oriented along the direction of the pointer. This field, interacting with the permanent magnet, will attempt to align itself with the latter. However, the spring resists this alignment so that only a relatively small deflection of the needle (less than 90 degrees) occurs. The stronger the current passing through the coil, the stronger the magnetic field produced and the greater the deflection. If the current direction is reversed, so are the magnetic direction and the deflection of the pointer.

Figure 44 Cont.

B. An ammeter is simply a galvanometer with a shunt carrying most of the current past the movement, rather than through it. Due to the sensitivity of the parts, the galvanometer cannot carry currents much greater than a few microamps or milliamps. Consequently, in order to measure larger currents, it is necessary to design a shunt system of known resistance so that a specific fraction of the total load current passes through the movement. The scale is calibrated to indicate the actual load current producing the deflection through a small fraction of itself.

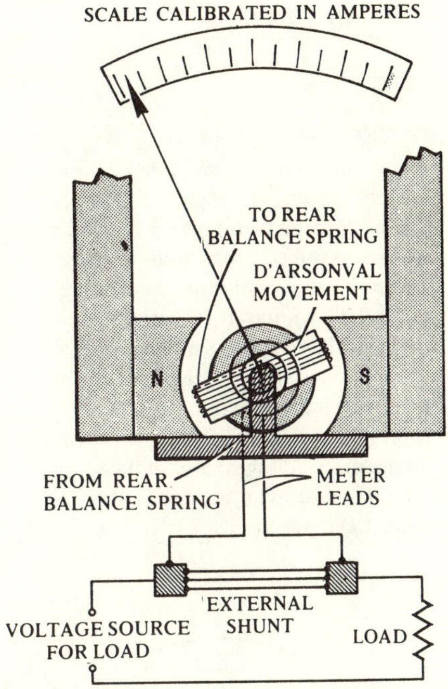

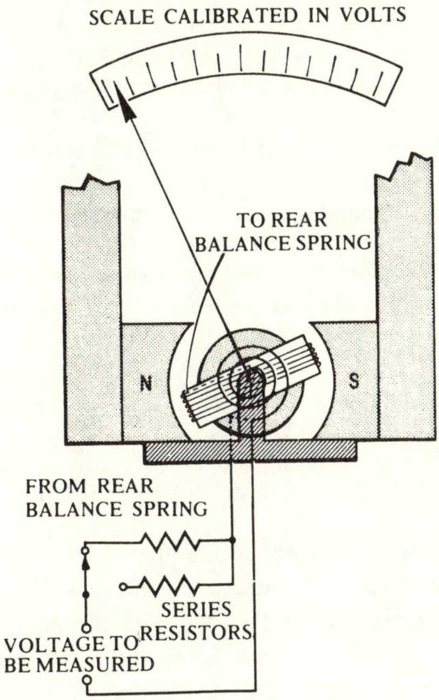

C. A voltmeter can also be constructed from the basic galvanometer. Since a known current causes a specific deflection of the pointer, Ohm's law (E=IR) can also be applied to translate the current (I) reading into a voltage (E). The resistance (R) is determined from the interval resistance of the meter and from applied series resistances, chosen to give the voltage ranges desired. For example, to measure a one volt potential difference with a full scale deflection on a galvanometer which needs $100\mu A (10^{-4} A)$ for such a deflection, $R = 1/10^{-4} = 10{,}000$ ohms for the combined meter and series resistance.

Figure 44 *Cont.*

D. The revolving field AC generator is the most widely used type. Direct current from a separate source passes through windings on the rotor by means of sliprings and brushes, maintaining a rotating electromagnetic field similar to a rotating bar magnet. The rotating magnetic field extends outward through the armature windings imbedded in the surrounding stator. As the rotor turns, alternating voltages are induced in the windings. The output is tapped from fixed terminals T1 and T2.

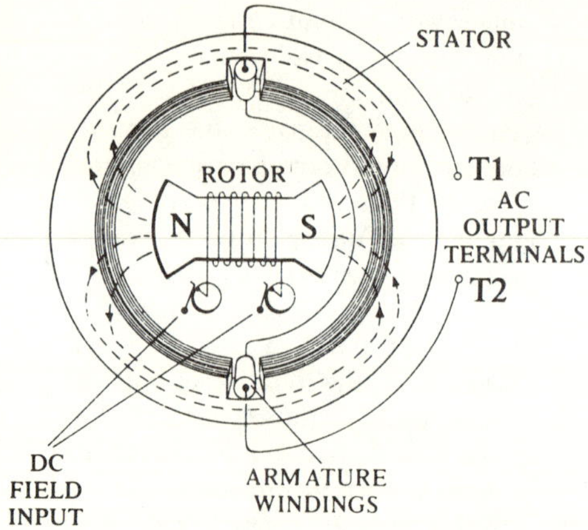

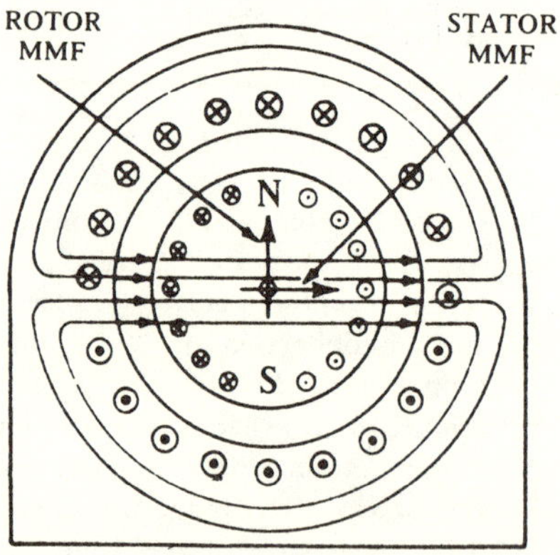

E. An AC motor is identical in principle to an AC generator run backwards. Instead of tapping power from the terminals, it is fed in, creating a magnetic field in the stator windings. This field interacts with the field of the rotor causing an alignment (rotation). Since alternating current is used, the direction of the stator field is continuously changing, so the alignment of the rotor is actually a continuous rotation. The rotor has a shaft attached so that mechanical energy can be extracted from the motor.

$$\mathbf{curl\ E} = \frac{1}{c}\ \frac{-\partial \mathbf{B}}{\partial t}.$$

The relationships involved are quite simple. When the current is increased in the primary coil, a current will be generated in the secondary coil proportional to the rate of increase, but in the opposite direction. When the current is reduced in the primary, current will be produced proportionally in the secondary, but this time in the same direction as in

Figure 45

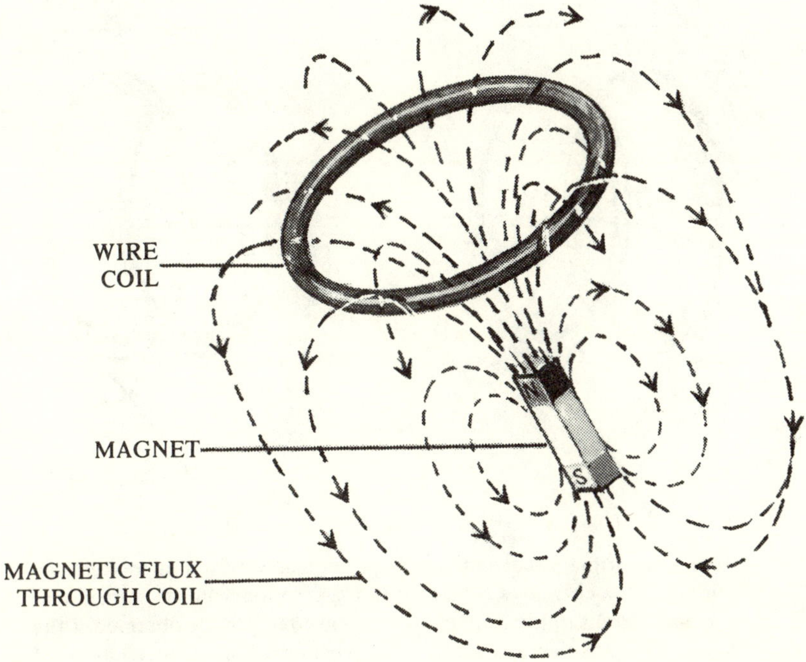

Faraday's induction law. As either the magnet or the wire coil are moved, the "number of field lines" enclosed by the coil changes. This change in the flux through the coil "produces" an electromotive force, hence a current in the wire.

the primary. Similarly, when the primary coil through which current is flowing is brought toward the secondary coil, a current is produced in the secondary proportional to the motion of the primary, but in the opposite direction. [Figure 46]

There is an amusing statement of this interrelation known as Lenz's law: In case of a change in a magnetic system, that thing happens which tends to oppose the change.

A look at the electromagnetic field which contains such an interconnected system shows that this interplay of the two currents guarantees a metastability which would be violated if, for example, current were produced in the same direction in a secondary wire upon its approach to a

Figure 46

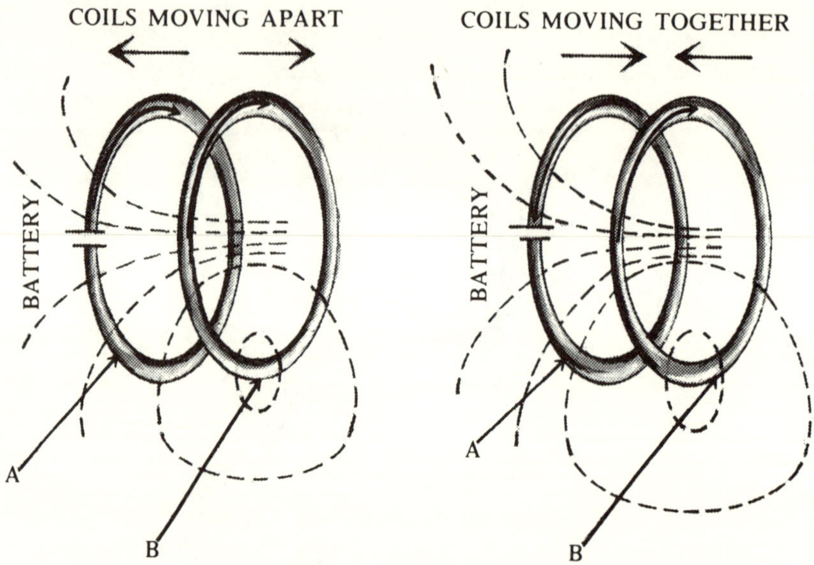

A. When the primary (current-carrying) coil (A) is moved away from the secondary (B), a current is generated in the secondary in the same direction as the current in the primary. The same effect will be observed if the primary current is decreased and if the secondary coil is moved away. If the coils are moved together or if the current in the primary is increased, the induced current in the secondary coil will flow in the opposite direction to that in the primary. It can be shown from energy considerations that if the current directions were opposite to what is observed, energy would be created *ex nihilo* in the induction process.

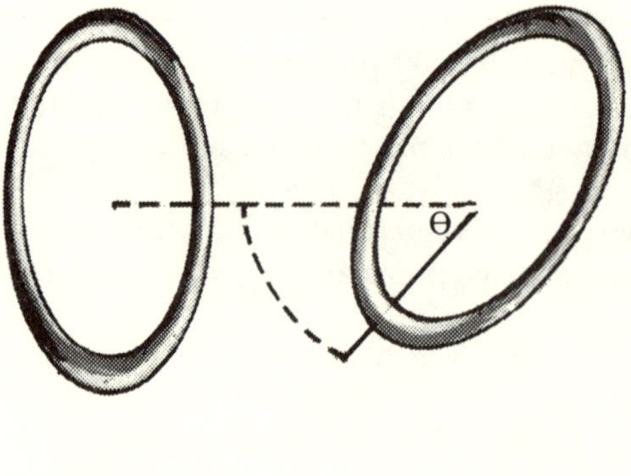

B. The amount of current induced in the secondary coil depends on a number of geometric factors. If either the primary or secondary coil is doubled into two loops, the induced current will be approximately doubled. If the angle between the coils is varied so that their surfaces are not parallel, the induced current will vary as the sine of the angle.

primary. Were that the case, the law of conversion of energy would be violated in a way not to produce a higher negentropic order, but merely a tendency to chaos and instability. In short, at the very least, the geometry of the universe would be other than we now know it.

Consider two coils, one bearing a primary current. Accelerate that current and a small transient current is produced in the secondary wire. This transient current is running in an opposite direction to the primary and the wires experience a transient ponderomotive force tending to separate them. This force, being in the direction of increasing the distance between them, will generate a current in the same direction in the secondary wire, creating an infinitesimal tendency toward the primary, and so on. Meanwhile, consider the primary wire after current has been generated in the secondary, causing it to move away from the primary. An extremely minute current will be generated in the primary by the secondary. This current will oppose the current already flowing through it (the reader should figure out why), thus slightly reducing the amount of original current. The quantities involved will be second order, but the interplay is like that of the pendulum in motion in respect to the relationship, although the effect is too small to effect an appreciable displacement. Were things to work oppositely to Lenz's law so that, for example, drawing two wires together generated current in the same direction in each wire, energy would be generated *ex nihilo* and the universe would be fundamentally unstable.

We have now almost completed the development of Maxwell's equations. But first, it is necessary to consider the expression for **curl E**. This is equal to the rate of change of the magnetic force $\partial \mathbf{B}/\partial t$ and it will not be zero when current is generated. Therefore, **E** cannot be the gradient of an electrostatic potential function. It is reasonable to consider a potential function such that

$$\mathbf{E}=-\mathbf{grad}\, V-\frac{1}{c}\frac{\partial \mathbf{A}}{\partial t}$$

The derivation rests on the fact that **curl A=B**; **A**, it should be recalled, equals

$$i\int \frac{\mathbf{ds}}{r}.$$

Therefore,

$$\int \mathbf{E}\cdot\mathbf{dl}=-\frac{1}{c}\int \frac{\partial \mathbf{A}}{\partial t}\cdot\mathbf{dl},\quad \int\left(\mathbf{E}+\frac{1}{c}\frac{\partial \mathbf{A}}{\partial t}\right)\cdot\mathbf{dl}=0.$$

This, in fact, was exactly the formula for the electromotive force developed by Neumann, a collaborator of Riemann's. Neumann, Riemann, and Weber derived the equation for the generation of electricity by directly referring to the source, that is, the current. Faraday de-emphasized the source by taking the magnetic flux variation as the active agency for electric induction. In practice, most calculations of necessity relate back to the source, thus substantiating the validity of the Riemann approach.

Next, we wish to look at **curl B**. Is a magnetic field only produced around a current flowing through a closed current? Maxwell answered no and correctly hypothesized the production of a magnetic field by a varying electric field. He reified this notion by naming the variation, $\partial \mathbf{E}/\partial t$, the displacement current. A steady field **E** will maintain an approximately steady flow of current. If we consider any cross-section through the circuit, approximately the same number of electrons will pass through in any given time period. Suppose, however, that the current is interrupted by a capacitor. Electrons will be collected on one plate and discharged off the other. But, the current flow will still be maintained as the capacitor is charged. [Figure 47] The field between the plates will be increased. It is the rate of increase of this field which can be thought of as a pseudo-current.

Mathematically, the divergence of any curl function is always zero because of the commutativity of second derivatives,

$$\frac{\partial^2 \mathbf{B}}{\partial x\, \partial y} = \frac{\partial^2 \mathbf{B}}{\partial y\, \partial x}\cdot$$

Therefore, the equation **curl B**$=4\pi\mathbf{J}/c$, where **J** is the current density, was incomplete from strictly mathematical considerations. The divergence of **J**, in some cases, would not be equal to the divergence of **curl B**, which is zero. If we only consider the current **J**, at the approach to a capacitor, the current is interrupted at that point with the capacitor plate acting as an electron source or sink. Therefore, Maxwell introduced the displacement current **J**′ as a continuation of the current **J**.

$$\mathbf{J}' = \frac{1}{4\pi}\frac{\partial \mathrm{E}}{\partial t}$$

is known as a displacement current. The build-up of charge on the plates, which can be described as a nonzero divergence for the current density, is responsible for the existence of a displacement current in the gap between the plates. This implies the dependence of the displacement current on open circuitry, since, in a closed circuit, the divergence of the current density will be zero. In the case of both electrical induction and transmission of electric signals by radiation, generation occurs as a result of the acceleration of charged particles. However, radiation and induction are not identical. In the case of radiation, it is the second time derivative

Figure 47

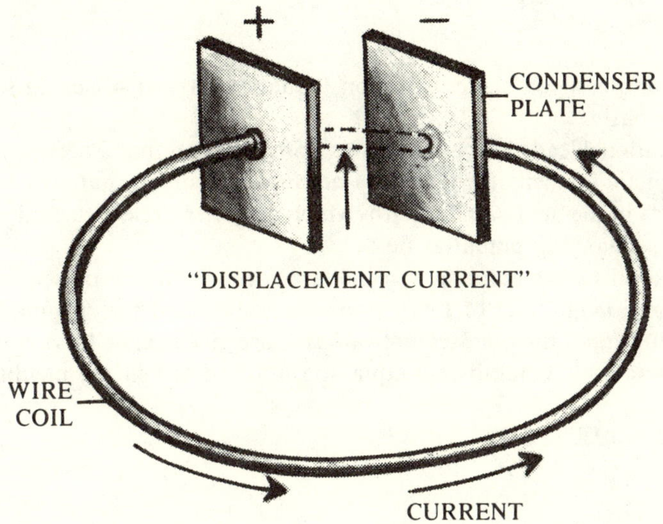

The so-called displacement current "flows" between the condenser plates.

change in the electric field which effects the pulse variation of the radiation, while, in the case of induction, the first time derivative change of the electric field is associated with the changing magnetic flux, hence with the induced electric field.

The time variation of electric displacement $\partial \mathbf{E} / \partial t$ is an essential factor in radiation. Maxwell's introduction of the notion of a "displacement current" as either a mathematical convenience or the imaginary current which closes an open circuit confuses an otherwise obvious and necessary concept. Maxwell reifies $\partial \mathbf{E} / \partial t$ in order to obtain a spurious justification for radiation as a kind of electrical generation occurring according to the rate of variation of the displacement current "in space." This is $\partial^2 \mathbf{E} / \partial t^2$.

The correct equation then is

$$\mathbf{curl\ B} = \frac{4\pi}{c}\mathbf{J} + \frac{1}{c}\frac{\partial \mathbf{E}}{\partial t}$$

Maxwell's equations can now be written in full:

1. $\mathbf{div\,E} = 4\pi\varrho$

2. $\mathbf{div\,B}=0$

3. $\mathbf{curl\ E}=-\frac{1}{c}\frac{\partial \mathbf{B}}{\partial t}$

4. $\mathbf{curl\ B}=\frac{4\pi}{c}\mathbf{J}+\frac{1}{c}\frac{\partial E}{\partial t}$

By simple algebraic substitution, equations 3 and 4 can be solved simultaneously.

Equation 4 leads to the interesting consideration that a nonconservative magnetic intensity (producing a nonzero curl upon rotation in a circuit) exists in the field even for paths which do not encircle a current coil if there is a changing potential field.

Maxwell's wave equations in modern form, with **E** representing an electrostatic magnitude of force c times greater than the magnitude **B** which determines the ponderomotive force, are given below for a point in space where $\mathbf{J}=0$. Quantity c is equal to the speed of light in magnitude.

$$\frac{\partial^2 \mathbf{E}}{\partial x^2}+\frac{\partial^2 \mathbf{E}}{\partial y^2}+\frac{\partial^2 \mathbf{E}}{\partial z^2}=\frac{1}{c^2}\frac{\partial^2 \mathbf{E}}{\partial t^2}$$

and

$$\frac{\partial^2 \mathbf{B}}{\partial x^2}+\frac{\partial^2 \mathbf{B}}{\partial y^2}+\frac{\partial^2 \mathbf{B}}{\partial z^2}=\frac{1}{c^2}\frac{\partial^2 \mathbf{B}}{\partial t^2}\ .$$

A simple solution for these wave equations is $\mathbf{E}=\mathbf{E_0}\sin[t\text{-}(r/c)]$ and $\mathbf{B}=\mathbf{B_0}\sin[t\text{-}(r/c)]$.

$$\frac{\partial E}{\partial t}=\mathbf{E_0}\cos\,[t-(r/c)],$$

$$\frac{\partial^2 \mathbf{E}}{\partial t^2}=-\,\mathbf{E_0}\sin\,[t-(r/c)],$$

$$\frac{\partial \mathbf{E}}{\partial r}=-\,\frac{\mathbf{E_0}}{c}\cos\,[t-(r/c)],$$

$$\frac{\partial^2 \mathbf{E}}{\partial r^2}=-\,\frac{\mathbf{E_0}}{c^2}\sin\,[t-(r/c)],$$

$$\frac{\partial^2 \mathbf{E}}{\partial r^2}=\frac{\partial^2 \mathbf{E}}{\partial x^2}+\frac{\partial^2 \mathbf{E}}{\partial y^2}+\frac{\partial^2 \mathbf{E}}{\partial z^2}=\frac{1}{c^2}\frac{\partial^2 \mathbf{E}}{\partial t^2}\ .$$

[Figure 48]

It should be noted that the Maxwell wave equations written in terms of the electric and magnetic fields above can be cast into a form using the scalar and vector potentials which explicitly accounts for the existence of sources in the field.

Figure 48

The series of four experiments described here are generally considered the critical experiments which validate Maxwell's equations.

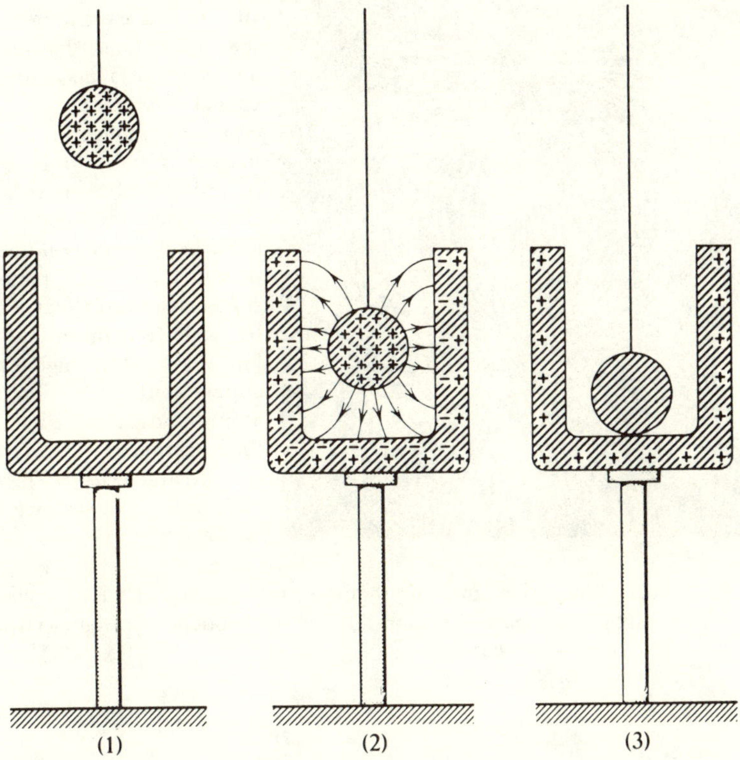

A. Gauss's law for electrostatics, which is equivalent to Coulomb's inverse square force law for static changes, can be written $\mathbf{divE} = 4\pi\varrho$. Benjamin Franklin first observed the experimental verification of this statement in 1775, although he did not entirely recognize its significance. In a letter to a friend, he wrote:

> I electrified a silver pint cann, on an electric stand, and then lowered into it a cork-ball, of about an inch diameter, hanging by a silk string, till the cork touched the bottom of the cann. The cork was not attracted to the inside of the cann as it would have been to the outside, and though it touched the bottom, yet when drawn out, it was not found to be electrified by that touch, as it would have been by touching the outside.

Figure 48 Cont.

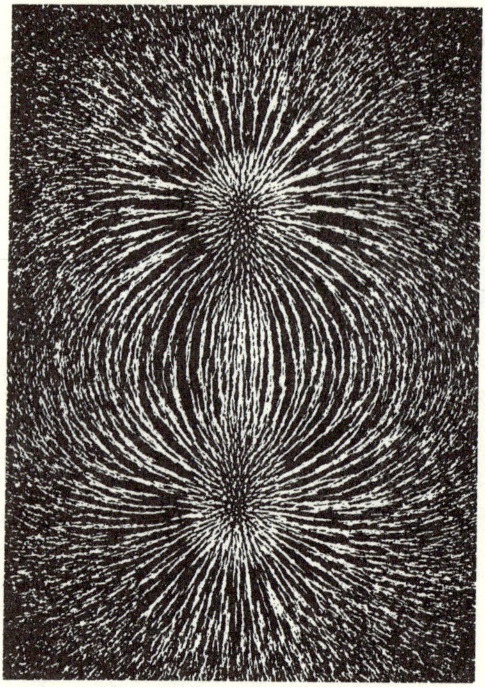

B. Gauss's law for magnetostatics, **div B**$=0$, is equivalent to the statement that no magnetic monopoles exist. In the electrostatic case, the net charge Q acts as a source or sink for the electric field. The law here states that sources and sinks of a magnetic field cannot be separated from each other. Consequently, if a bar magnet is broken in two, each piece has both a north and a south pole. Presumably, this could be done ***ad infinitum***, although there is no reason to state categorically that it must be so. In fact, during the past few years, some high energy experiments involving subatomic particles have indicated that perhaps this law is violated at that level. These experiments have not yet been confirmed.

C. Faraday's induction law states that the time variation of the magnetic flux in a region of space is responsible for the production of an electric field in that region, that is

$$\mathbf{curl\ E} = -\frac{1}{c}\frac{\partial \mathbf{B}}{\partial t}.$$

This effect is easily demonstrated by many examples discussed earlier in the text. Basically, a bar magnet caused to move near a coil of wire will generate a current in the wire. If a sufficiently tightly wound solenoid is used for this experiment, the resistance to the magnet's motion can be readily felt. Such a resistance is the evidence of work being done to convert mechanical into electrical energy.

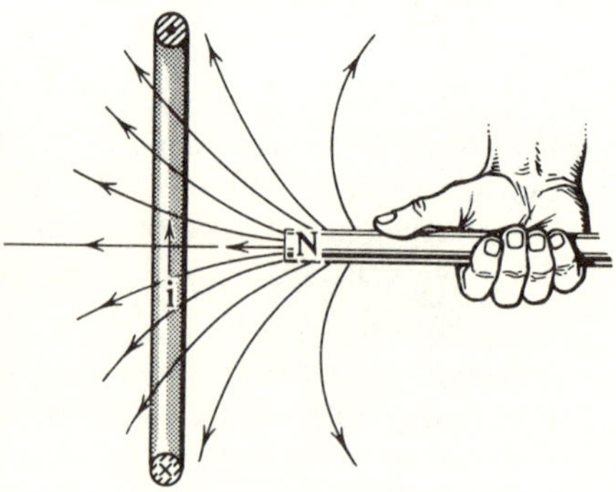

Figure 48 Cont.

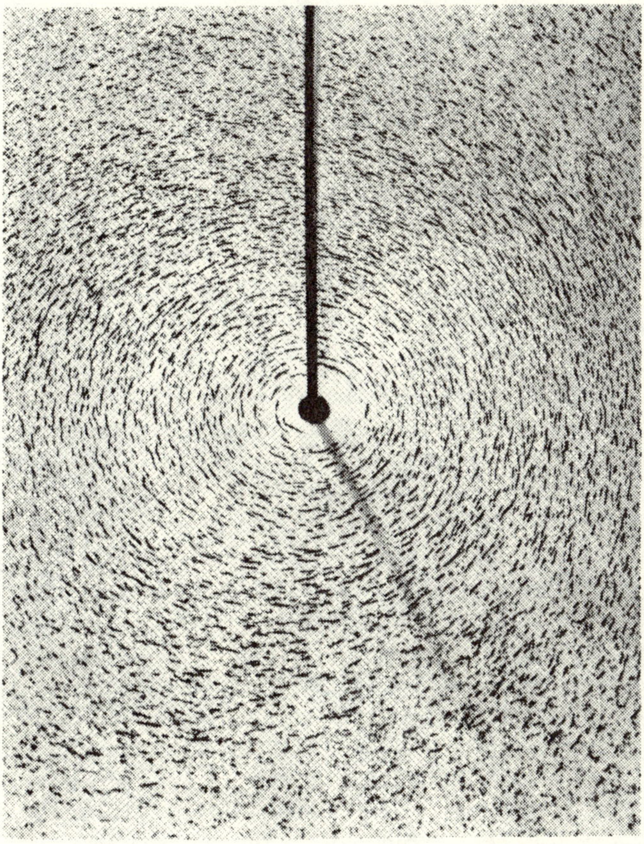

D. Ampère's law states that a current induces a magnetic field in the space surrounding it. Maxwell modified this statement to draw out the intrinsic symmetry between the electric and magnetic field, postulating that the time variation of an electric field could also produce a magnetic field, that is

$$\textbf{curl B} = \frac{1}{c}\left(4\pi\,\mathbf{J} + \frac{\partial \mathbf{E}}{\partial t}\right).$$

Ampère's contribution is easily demonstrated by the iron filing patterns generated around a current-carrying wire. However, Maxwell's modification is very difficult to observe directly. Combined with Faraday's law, this statement describes the propagation of electromagnetic radiation at the speed of light, a phenomenon which has repeatedly been confirmed throughout the electromagnetic spectrum.

$$\frac{\partial^2 V}{\partial x^2} + \frac{\partial^2 V}{\partial y^2} + \frac{\partial^2 V}{\partial z^2} - \frac{1}{c^2}\frac{\partial^2 V}{\partial t^2} = -4\pi\varrho,$$

$$\frac{\partial^2 A}{\partial x^2} + \frac{\partial^2 A}{\partial y^2} + \frac{\partial^2 A}{\partial z^2} - \frac{1}{c^2}\frac{\partial^2 A}{\partial t^2} = \frac{4\pi}{c}\mathbf{J}.$$

The quantities **E**, **H**, and **B**, used by Maxwell, represent forces. Therefore, as the argument of this book has developed, they are derived, rather than primary. It is to the potential that we must look to obtain a more adequate approximate understanding of the electromagnetic field. Not only Riemann, but a contemporary of his, Lorenz (not to be confused with Lorentz), also developed the notion of retarded potential from which Maxwell's forces can be derived. Maxwell was familiar with his published work, but disregarded it. We quote Lorenz's excellent explanation of the retarded potential.

> As the laws of induced currents generally admitted and based on experiment did not lead to the expected result, the question was whether it was not possible so to modify the laws assumed that they would embrace both the experiments on which they rest and the phenomena which belong to the theory of light. . . . It is at once obvious that the equations, which are deduced in a purely empirical manner, are not necessarily the exact expression of the actual law; and it will always by permissible to add several members or to give the equations another form, always provided these changes acquire no perceptible influence on the results which are established by experiment. We shall begin by considering the two members on the right side of the equations (i.e. V_E and A — C.W.) as the first members of a series. . . . These equations . . . express further that the entire action between the free electricity and the electrical currents requires time to propagate itself — an assumption not strange in science and which may in itself be assumed to have a certain degree of probability. For, in acccordance with the formulae found, the action in the point (x y z) at the moment t does not depend on the simultaneous condition in the point (x′ y′ z′), but on the condition in which it was at the moment t-(r/c); that is, so much time in advance as is required to traverse the distance r with the constant velocity c.
>
> The theoretically important conclusion would thence follow . . . that electrical forces require time to travel, and that every action of electricity and of electrical currents does in fact only depend on the electrical condition of the immediately surrounding elements.

Riemann's Equations

Riemann did not explicitly set down the equation for the retarded potential A, although he did note the relationship

$$\frac{1}{c}\frac{\partial V_E}{\partial t} = \mathbf{grad\ A}.$$

Riemann took Neumann's equation for the induction of electromotive force

$$\left(\frac{d\mathbf{A}}{dt}\cdot \mathbf{ds}\right)$$

and used it as the basis for redefining the notion of potential energy to include the potential for electrons in motion to induce electricity. Reasoning backward from the formula representing the induction of electricity in two current-bearing coils because of their mutual variation, he arrived at a velocity potential.

He broke the total work into three components. First, the electromotive work in the first conductor:

$$dt\int ds\left(i_1\frac{\partial v_1'}{\partial t}+i_2\frac{\partial v_2'}{\partial t}+i_3\frac{\partial v_3'}{\partial t}\right).$$

where v_1', etc. are the velocities of the electrons in the x, y, and z directions. Similiarly, i_1 represents the component of current in the first wire which is in the x direction; dt is a given time period. Second, the electromotive work in the second wire:

$$dt\int ds'\left(i_1'\frac{\partial v_1}{\partial t}+i_2'\frac{\partial v_2}{\partial t}+i_3'\frac{\partial v_3}{\partial t}\right).$$

Third, the electrodynamic work of both currents on each other:

$$dt\int \mathbf{ds}\times\mathbf{ds}'\frac{[\partial(1/r)]}{\partial t}(i_1i_1'+i_2i_2'+i_3i_3').$$

In the first two cases, the current varies. In the third, it is the distance between the two which varies.

From this, Riemann deduces his expanded velocity potential which he calls D.

$$D=\int ds\,(v_1'i_1+v_2'i_2+v_3'i_3)+\int ds'\,(v_1i_1'+v_2i_2'+v_3i_3')$$

$$+\int\frac{ds\,ds'}{r}(i_1i_1'+i_2i_2'+i_3i_3').$$

This velocity potential subsumes the electrostatic potential in a relativistic treatment of electron velocity, correcting the former to anticipate Lorentz's later work. (Riemann's development of this material appears in Part II of this book.)

The equations for D can be transformed to a form which only depends upon the relative velocities of the electrons whose flow creates the two currents:

$$D = \frac{1}{c^2} \sum \frac{Q_1 Q_2}{r} \{(v_1 - v_1')^2 + (v_2 - v_2')^2 + (v_3 - v_3')^2\}$$

When this is added to the electrostatic potential,

$$D = \sum \frac{Q_1 Q_2}{r} \left(1 - \frac{v^2}{c^2}\right).$$

Because of its importance in establishing that Riemann did, in fact, anticipate relativity theory, we will quote Lorentz's development in *The Theory of Electrons* (minor stylistic changes are made for uniformity of notation — C.W.):

> We may therefore, without too abrupt transition, turn once more to some questions belonging to what we can call the dynamics of an electron, and in which we are concerned with the field the particle produces and the force exerted on it by the aether. We shall, in this way, be led to the important subject of the electromagnetic mass of the electrons.
>
> To begin with, I shall say some words about the field of a system of electrons or of charges distributed in any way, having a constant velocity of translation v, say in the direction of the axis of x, smaller than the speed of light c. We shall introduce axes of coordinates moving with the system. Now, we have already seen that the field is carried along by the system. The same may be said of the potentials of V and **A**, which serve to determine it, and it may easily be inferred from this that the values of $\partial V/\partial t$ and $\partial \mathbf{A}/\partial t$ in a fixed point of space are given by
>
> $$-v\frac{\partial V_E}{\partial x}, \quad -v\frac{\partial \mathbf{A}}{\partial x}.$$
>
> Similarly,
>
> $$\frac{\partial^2 V_E}{\partial t^2} = v^2 \frac{\partial^2 V_E}{\partial x^2}, \qquad \frac{\partial^2 \mathbf{A}}{\partial t^2} = v^2 \frac{\partial^2 \mathbf{A}}{\partial x^2}.$$
>
> Thus, equation

$$\frac{\partial^2 V}{\partial x^2} + \frac{\partial^2 V}{\partial y^2} + \frac{\partial^2 V}{\partial z^2} - \frac{1}{c^2}\frac{\partial^2 V}{\partial t^2} = -Q$$

takes the form

$$\left(1 - \frac{v^2}{c^2}\right)\left(\frac{\partial^2 V}{\partial x^2} + \frac{\partial^2 V}{\partial y^2} + \frac{\partial^2 V}{\partial z^2}\right) = -Q$$

and, similarly,

$$\left(1 - \frac{v^2}{c^2}\right)\left(\frac{\partial^2 A_x}{\partial x^2} + \frac{\partial^2 A_x}{\partial y^2} + \frac{\partial^2 A_x}{\partial z^2}\right) = -\frac{v}{c}Q.$$

Comparing these, we conclude that

$$A_x = \frac{v}{c}Q.$$

This can be effected by a suitable change of independent variables. If a new variable x' is defined by

$$x' = \frac{x}{\sqrt{1 - (v^2/c^2)}},$$

then we have

$$\frac{\partial^2 V}{\partial x^2} + \frac{\partial^2 V}{\partial y^2} + \frac{\partial^2 V}{\partial z^2} = -Q,$$

having the well-known form of Poisson's equation. Since this equation occurs in the determination of the field for charges that are at rest, the problem is hereby reduced to an ordinary problem of electrostatics. Only the value of V_E in our moving system S is connected with the potential, not of the same system when at rest, but of a system in which all the coordinates parallel to OX have been changed in the ratio determined by

$$x' = \frac{x}{\sqrt{1 - (v^2/c^2)}}.$$

The result may be expressed as follows. Let S′ be a system having no translation, and which we obtain by enlarging the dimensions of S in the direction of OX in the ratio of

$$\frac{1}{\sqrt{1 - (v^2/c^2)}};$$

then, if a point with the coordinates x, y, z in S and a point with the coordinates x', y, z in S′ are said to correspond to each other, if the charges of corresponding elements of volume are supposed to be equal, and if V′ is the potential in S′, the scalar potential in the moving system is given by

$$V = \frac{V'}{\sqrt{1-(v^2/c^2)}}.$$

Therefore, in Lorentz's system V′ is comparable to Riemann's D. Riemann, of course, did not make the relativistic correction of length which followed from the Michelson and Morley experiments.

Riemann rigorously derived electrostatics as a subsumed feature of electrodynamics. He deduced the behavior of a charged particle from that of a current, in this case showing that the relative velocity of two charges, rather than an "absolute" velocity, is a determining factor in their interaction. The normal reductionist interpretation of current is to treat it as the sum of motions of a number of charges. On this basis, we would expect two electrons moving at the same velocity to have an attractive interaction as do two current elements. The Trouton and Noble experiment directly contradicts this expectation, showing that the current can not be treated in the reductionist manner, but, in fact, represents a higher ordering of the field.

The Lorentz Force Law

Lorentz adduced his force law, which describes the interaction of charged matter with a magnetic field, from the electron law of Clausius. While it is only Riemann and, later, Lorentz who formulated an electron law in terms of retarded potentials, Weber also had an electron law. The work of the four men can be considered to be broadly along the same lines. Riemann modified Weber's law which was cast in terms of relative velocities, but only as these were projected radially. Riemann's law was explicitly relativistic in its final form. Clausius and Lorentz treated the velocities of the two currents as absolute velocities with reference to a third coordinate system. The elaboration necessary in the theory of relativity, which transforms the coordinate system so that it travels with one of the currents, is merely a complicated way of conceptualizing relative velocity.

Lorentz took Clausius's formula for the kinetic potential of two electrons where A is given as an ordinary potential function and transformed it into a retarded potential, so that

$$\frac{dA_x}{dt} = \left(\frac{\partial A_x}{\partial t} + \frac{\partial A_x}{\partial x} v_1 + \frac{\partial A_x}{\partial y} v_2 + \frac{\partial A_x}{\partial z} v_3 \right)$$

and

$$e\left(A_x\frac{v}{c} + A_y\frac{v}{c} + A_z\frac{v}{c} + V_E\right)$$

became

$$e\left(-\frac{\partial V_E}{\partial x} - \frac{1}{c}\frac{\partial A_x}{\partial t}\right) + \frac{e}{c}v_3\left(\frac{\partial A_z}{\partial x} - \frac{\partial A_x}{\partial z}\right) + \frac{e}{c}v_2\left(\frac{\partial A_y}{\partial x} - \frac{\partial A_x}{\partial y}\right).$$

$$e\left(\frac{\partial V_E}{\partial x} - \frac{1}{c}\frac{\partial A_x}{\partial t}\right)$$

is E_x (the x component of **E**).

$$\left(\frac{\partial A_z}{\partial x} - \frac{\partial A_x}{\partial z}\right) = \textbf{curl}\ \mathbf{A} = B_y.$$

The Lorentz force law then is

$$e\mathbf{E} + \frac{e}{c}(\mathbf{v}\times\mathbf{B}).$$

The way Lorentz's force law is taught today obscures its origins in potential theory to the point where it is virtually mystified. Witness the following "definition" of magnetic induction:

> For the purpose of defining the magnetic induction, it is convenient to define F_m the magnetic force (frequently called the Lorentz force) as that part of the force exerted on a moving charge which is neither electrostatic nor mechanical. The magnetic induction, **B**, is then defined as the vector which satisfied $F_m = Q\mathbf{v}\times\mathbf{B}$ for all velocities.

This does double duty as a definition of Lorentz force law. The quotation is from *Foundations of Electromagnetic Theory,* in some respects a usefully rigorous text.

Lorentz's force law has the advantage of describing the path of a moving electron as it is affected by a magnetic field. This focus upon the mediation of the direct action of one electron on another through the field. makes a bridge between the older electron theories and Maxwell's theory. Lorentz's own treatment of his work is exemplary; however, it is very much prey to the kind of formalism cited above when it is applied by less scrupulous writers.

The Lorentz law accomplishes a nice unification between the ponderomotive force and the generation of current which occurs when two currents approach each other. The example of a cyclotron shows this. The magnet of a cyclotron provides a region of constant magnetic force. If a charge starts to move in the direction of the field B, there will be no force and, hence, the particle will continue to move in this direction without acceleration. If the particle is started in a plane perpendicular to B, its acceleration will lie in the same plane and the particle's path will be contained in this plane. Further, since the force is at right angles to the velocity, no work will be done and the speed will remain constant. Thus, as the particle moves, it will change its direction under the action of a constant force at right angles to its path. It will, in short, move in a circle. [Figure 49] We can then consider the motion of two currents toward each other under the action of the ponderomotive force as a tendency of the current to turn in that direction which is constrained by the material properties of the conductor. The generation of current, as a result of the motion of the conductors which are drawn together by the ponderomotive force can also be described by the Lorentz force operating on the electrons in the conductor as they move toward each other. Equally, we can look at it as the conversion of energy from an electrodynamic force to an electromotive force. In neither interpretation is new energy introduced into the system.

The Screw Field

It seems very likely that the Lorentz force law will not prove to be generally valid. Electrons and positive ions in high energy plasmas, subjected to magnetic fields in plasma focus machines, have been observed to "violate" the Lorentz force law as they are in the process of forming themselves into vortical filaments. These energy dense plasmas appear to incorporate their own magnetic field and the external field in such a way as to create extreme currents of concentrated energy internally. Ion and electron flows within these vortices do not circulate around magnetic field lines, but, instead, form filaments which flow along paths which change from being virtually circular at the outside perimeter to vertical at the center. Magnetic field lines are "frozen into" the plasma flow so that the path of the flow and field lines are in the same direction. [Figure 50]

This "unlawful" behavior of the plasmas was anticipated in an interesting way by Felix Klein who pointed out the analogy between the electromagnetic field and a screw field. The motion of a rigid body, being composed of a rectilinear translation and a rotation whose axis is in the direction of the translation, effectively the motion of a screw, is generated

Figure 49

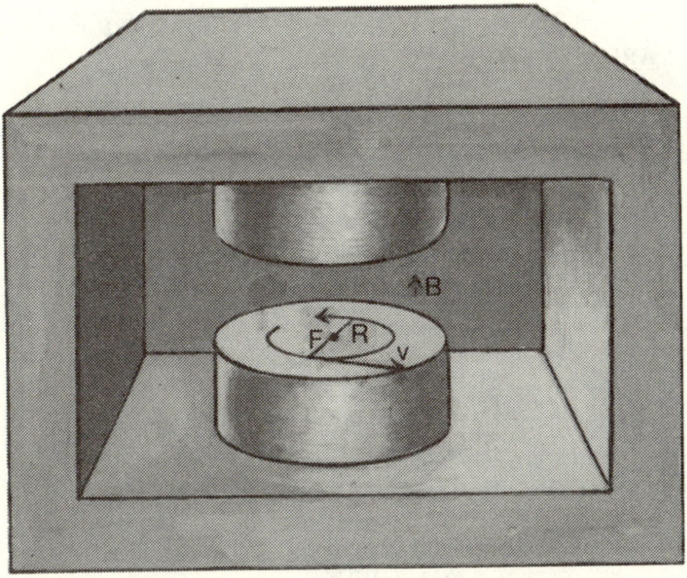

A uniform magnetic field is created between the magnet poles in the cyclotron, where a charged particle will move in a circular path, due to the force generated, normal to its velocity and to the magnetic field. The magnetic force (qvB) is balanced by the çentrifugal forcę (mv^2/R), resulting in a circular path with radius $R=mv/qB$, where m is the particle mass and q is its charge.

---

by the combination of a force and a rotary couple. An electric field imparts the rectilinear motion to a charged particle, while a magnetic field provides the rotary motion. In the event that a magnetic field is held constant and a stream of electrons is unconstrained, it will follow a helical path around the field lines, depending upon its initial velocity. [Figure 51]

Although Klein suggested that the electromagnetic field is a screw field, to our knowledge he did not elaborate this idea. Such a screw field would seem to offer a possibility for describing the behavior of plasmas in terms of Lorentz's law, but that in itself would be anomalous in our opinion. Nonetheless, the screw field does suggest a notion of varying force-free planes within the typical electromagnetic field.

An important feature of the screw field model is the determination of

Figure 50

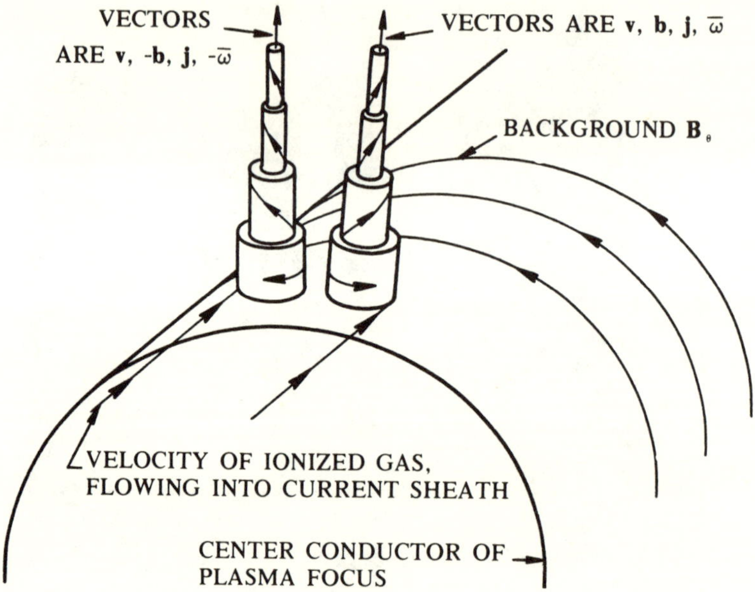

Dissected diagram of the vector configuration of a pair of paramagnetic vortex filaments formed in the current sheath of the plasma focus; **v** is the mass flow velocity, **b** is local magnetic field, **j** is current density, $\overline{\omega}$ is vorticity, $\mathbf{B}_\theta$ is background magnetic field caused by flow of current in the coaxial electrodes. The behavior of the vector quantities in the filaments bears a striking resemblance to the behavior of null planes in Klein's screw field formulation.

---

such rotating force-free planes which are themselves slanted like the threads of a screw. These slanted null planes are determined by the pitch of imaginary screws of increasing circumference, which surround the axis at greater and greater radial distances such that it takes a turn through an increasingly larger arc length to effect a vertical motion of the screw. The pitch of such a screw will decrease with radial distance from the axis. The analogy to the situation in the plasma filaments, [Figure 50] where the vector quantities are all pointed parallel to each other with a pitch decreasing with increasing distance from the center of the filament is quite striking. A more precise determination depends on experiments and

Figure 51

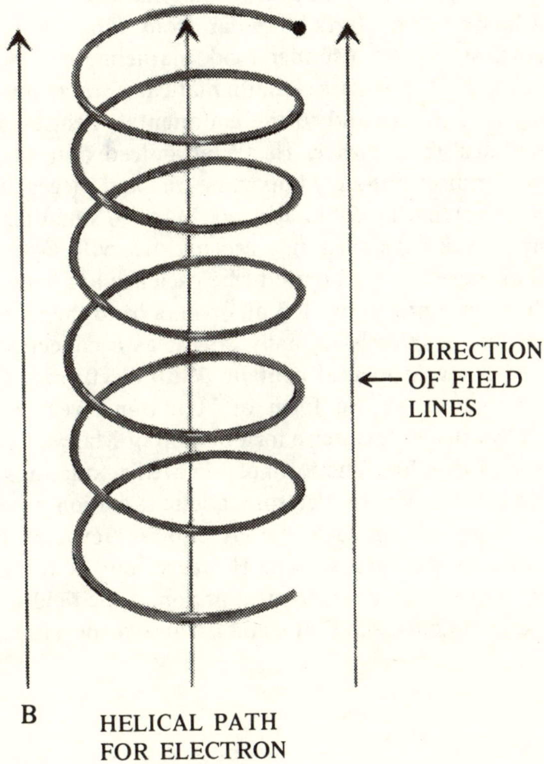

An electron, entering a region of uniform magnetic field, will tend to move in a helical path. The particle velocity can be broken down into a component perpendicular to the magnetic field and another parallel to it. It is the perpendicular component which contributes to the circular motion (see Figure 49), while the parallel component provides for a vertical drift.

---

detailed calculation of what occurs in the filaments in light of Klein's idea. Do these fields violate Maxwell's laws? It is yet to be determined.

Mathematics is not epistemologically neutral. When it is treated as such, it is merely the algebraic residue, the detritus of a determining geometry. The mathematician who ascribes such neutrality to an algebraic formalism endorses an epistemology — nominalism. The geometry of such algebraicism coheres with the world view of monetarism. The world is made up of a collection of discrete things which are there for the grabbing. Lawfulness is the arbitrary rule of those in power. Such a world view must hysterically deny the reality of progress and reduce geometry to a parody of Euclidean axiomatics.

Maxwell and Hertz were far from subscribing to such atrocities. Nonetheless, both misunderstood mathematics and physics. Maxwell supposed that the same mathematical formalisms could be applied equally to Faraday and to the continental theories of electromagnetism without distorting either. Hertz recognized that this led in practice to Maxwell indiscriminately importing bits and pieces of both, but thought, nonetheless, that he could abstract Maxwell's mathematics from his confusion: "Maxwell's formulae are his theory."

This heritage, in a period when science has been under severe attack for almost a century, has left an open door to the worst sort of nominalist tendencies. Science is virtually taught as a collection of algebraic rules stripped of experimental content. With electromagnetism, this is particularly so because the form of Hamilton's vector notation allows an especially attractive concise formulation of Maxwell's wave equations and Lorentz's force law. The form of Maxwell's equations is adequate to describing the behavior of electromagnetic radiation waves in a field without charge; they are, in fact, merely another form of Riemann's retarded potentials as we have shown. However, out of the context of potential theory, they import a mechanical notion of the field — or aether — which is at best a detour from that standard already developed by Riemann.

CHAPTER VI

The Theory Of Relativity

Typically, Einstein's Theory of Relativity is presented as the farthest extension of Maxwellian "classical" electromagnetic theory. In reality, the theory is a banalization of those first approximations which Riemann used as heurisms upon which to base his actual theoretical achievements. Einstein's General Theory of Relativity uses the first approximation curvature tensor derived in Chapter IV. In attempting to unify the gravitational and electromagnetic fields, Einstein overlooked exactly that notion of the increasingly complex topologies of higher-order energy potentiation which is the subject of Riemann's life work. Einstein's unified field explicitly denies the notion of a self-developing universe. His "expanding" universe is necessarily doomed to a catastrophic end, because he imposes on it the false condition of the conservation of energy.

It is useful to conclude this work by tracing the interplay between developing notions of light radiation and electromagnetic radiation — if for no other reason than that the establishment of this coherence dominated the thought of leading nineteenth century scientists. Relativity theory as expressed by Einstein, even though it banalizes the fundamental discoveries of Riemann, makes one nice point. Not only is the particle properly subsumed by the particle-field collectivity, but its so-called discrete existence, as measured in space and time, is subsumed by the motion of that collectivity.

Finally, Einstein's axiomatic assumption of the constant velocity of light properly exphasizes that the throughput of radiant energy has been, to date, the determining factor in the evolution of the biosphere.

The theory marked the culmination of nearly a century's work which sought to explain the failure of any experiment to detect those changes in the velocity of propagation of light which would be expected according to either ballistic or wave theories of light generation. If light is, in effect, a beam of corpuscles emitted from a source, then, like a ball thrown in a moving train, it should share in the velocity of its source according to the principle of inertia, but it does not. If light is transmitted as a wave impulse, then it must be affected by the interaction of the receiver and the medium. Yet, neither is this the case.

It is, in fact, the case that metrical relationships within any local section of the electromagnetic field are adjusted to the flow of energy into or out of the field. This is marked by the apparent slowing of time and shrinking of objects as a function of the measured velocity of a system. Any method which we may suggest to determine variations in the speed of light ultimately falls afoul of that series of interlocking relationships which are subsumed under the constancy of the ratio of the electromagnetic to the electrostatic unit of measure by which the speed of light is determined. The electrostatic unit measures the force exerted by two charged particles "at rest"; the electromagnetic unit measures the additional force exerted between two collections of these particles in motion. The ratio, is of course, c. Commonly, **E** is measured in electrostatic units, **B** in electromagnetic units. This being so,

$$\mathbf{curl\ E} = -\frac{1}{c}\frac{\partial \mathbf{B}}{\partial t}\,.$$

We have already contrasted relativity theory, as it was first developed by H.A. Lorentz, to its anticipation by Riemann. It is appropriate to locate the theory in the work of Lorentz, since it was his formulations which gave it shape. To Einstein goes credit for understanding that the theory demanded a new axiomatic foundation for physical geometry which would finally replace the Newtonian conception of "absolute" space and "absolute" time as an empty container in which to place geometry and mechanics. It was a bitter tragedy for the man and unfortunate for the rest of us that Bertrand Russell's gang used Einstein's significant contributions to the theory to peddle deliberately pernicious nonsense about cultural relativity — as if a standard for progress rests upon the length of the measure of a second or an hour, or the establishment of the simultaneity of events!

The gist of the axiomatic superstructure of the Special Theory of Relativity rests on the interpretations of the results of Michelson and Morley's experiments to determine the velocity of the earth relative to the velocity of the aether, the medium through which, it was presumed, light was transmitted as a wave impulse. As Maxwell first remarked, the time required for a ray of light to travel from a point A to a point B and back to A must vary by a small magnitude — of the second order — when the two points together undergo a displacement with respect to the aether. The Michelson experiment of 1881 and a refinement of it done by Michelson and Morley in 1887 discovered no such effect.

The theory which they tested asserted that the aether was at rest relative to the earth. It is clear that any geocentric theory which assumed that the aether traveled with exactly the velocity of the earth throughout the entire universe could only be acceptable to such absolutists as otherwise favored the Hapsburg Inquisition. It was Thomas Young and Augustin Fresnel who first raised the question of the relationship of the velocity of the aether to the velocity of the earth. It is necessary to bear in mind that the term "aether" implies the attempt to conceptualize the electromagnetic field. Initially questions about the drift of the aether were not raised to dispose of uncomfortable questions about the measurability of light. On the contrary, light was the appropriate means by which these scientists hoped to be able to measure and, therefore, determine the behavior of the field.

Wave Motion in the Aether

In 1804, Young took up the question of the aberration of the stars. The question which he sought to answer was why the aberration effect occurs if light is propagated through a medium as a wave motion rather than as a ray-like beam of particles. He wrote:

> Upon considering the phenomena of the aberration of the stars, I am disposed to believe that the luminiferous aether pervades the substance of all material bodies with little or no resistance, as freely perhaps as the wind passes through a grove of trees. If we suppose the aether surrounding the earth to be at rest and unaffected by the earth's motion, the light waves will not partake of the motion of the telescope, which we may suppose directed to the true location of the star, and the image of the star will therefore be displaced from the central spider-line at the focus by a distance equal to that which the earth describes while the light is traveling through the telescope. This agrees with what is actually observed.

This solution raises as many questions as it appears to answer. Suppose that in our telescope a slab of glass with a plane face is carried along by the motion of the earth. Should this glass be positioned at right angles to the true direction of the star after we have corrected for the aberration effect, or should it be placed with reference to the apparent direction of the star? The effect in question is refraction — a light beam going from one medium through another will be bent if it approaches the boundary of the two media at an oblique angle. This was submitted to a test by François Arago, who concluded that the light coming from any star behaves, in all cases of reflection and refraction, precisely as it would if the star were situated in the place which it appears to occupy in consequence of aberration, and if the earth were at rest. The apparent refraction in a moving prism is equal to the absolute refraction in a fixed prism. [Figure 52]

Fresnel now set out to provide a theory capable of explaining Arago's result. He adopted a suggestion of Young's that the refractive powers of transparent bodies depend upon the concentration of aether within them. He also assumed that when a body is in motion part of the aether within it, whose density is in excess of the density of "free" aether in a vacuum, is carried along by the motion of the body. This theory and the mathematical formulae which accompanied it were able to account for all effects observable at that time; these were in the first order range of magnitude.

A kind of experimental verification for the general line of reasoning was provided by Hoek in 1868. A beam of light was divided into two portions, one of which was made to pass through a tube of water, then reflected to a mirror, and afterwards allowed to return to its point of origin without passing through the water. The other portion of the bifurcated beam was made to describe the same path in reverse order so that it passed through the water on its return journey. On causing the two portions of the beam to interfere, no variation in the interference pattern was found despite the orientation of the apparatus with regard to the earth's motion.

Let w denote the velocity of the earth which we suppose to be directed from the tube towards the mirror. Let c/u denote the velocity of light in the water at rest and c/u + v the velocity of the light in the water when it is moving. Let l be the length of the tube and suppose the distance from the end of the tube to the mirror to be zero since it does not affect the outcome of the experiment, then the time taken by the first portion of the beam to perform its journey is evidently

$$\frac{l}{(c/u)+v-w}+\frac{l}{c+w},$$

while the time for the second trip is

$$\frac{l}{c-w}+\frac{l}{(c/u)-v+w}.$$

By equating the two expressions, v is given as

$$(u^2-1)\frac{w}{c},$$

when terms of higher order (i.e., smaller by being exponential) than w/c are neglected; u is the refractive index of water so that c/u denotes the velocity c_1 of light when it passes through water which is at rest. Fresnel had accorded a refractive index to the aether in a body and Hoek's result accorded exactly with his hypothesis.

Figure 52

Refraction is explained by both the corpuscular and the wave theories. The observed refraction is independent of whether the interface is at rest or in motion relative to the light source.

A. A ray of light changes direction while passing from air to glass. According to the corpuscular theory, the ray must travel faster in the glass in order to achieve the observed bending. The observed refraction is explained using Fermat's principle in conjunction with Figure 7.

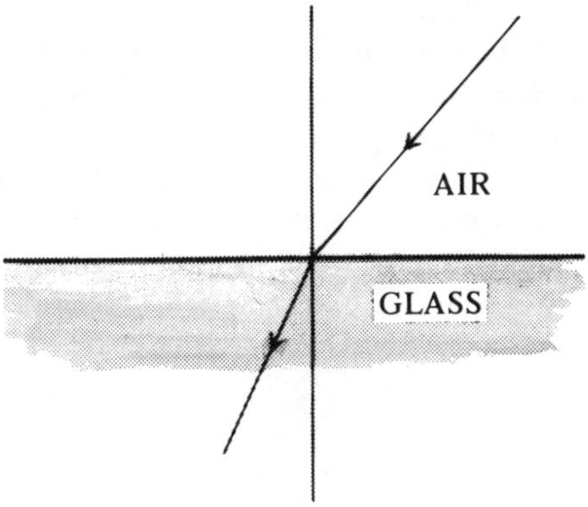

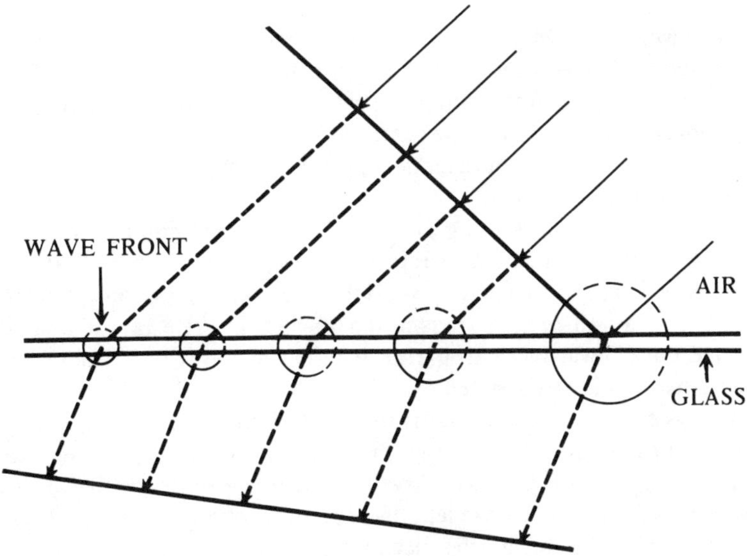

B. Huygens, a contemporary of Newton and friend of Spinoza, discovered the principle illustrated here. Every point of a surface on which the light impinges must be regarded as the source of a new spherical wave of light. The rays of the corpuscular theory become the direction of motion of the wave fronts. This theory agrees with the observed fact that the light travels more slowly in the glass.

The telescope, by which an emergent wave front is received, is itself being carried forward by the earth's motion. We must, therefore, make the usual correction for aberration in order to find the direction of a star. However, the aberration precisely counteracts those refraction effects which would be due to the motion of the glass medium. Therefore, the motion of the earth has no first-order influence on the refraction of light from the stars.

In 1845, Sir G.G. Stokes proposed a modification of Fresnel's theory to the effect that the earth's motion transmits motion to neighboring portions of the aether. This hypothesis was attractive because of the failure of experimentation to indicate any frictional effect as a result of the earth's abrasive contact with the aether, since, to account for its ability to propagate light waves with an enormous velocity, it was necessary to assume that the aether had some property akin to extreme elastic rigidity. In fact, any theory which imports material qualities to the aether must run into serious difficulties.

The Michelson-Morley Experiments

The Michelson-Morley experiments were designed to decide between the two hypotheses: Fresnel's or Stokes's. It was similar in form to Hoek's experiment except that the apparatus was refined to the point where any second-order differences in time of travel could also be measured. Michelson believed this to be a confirmation of Stokes's theory. The details of the experiment are as follows:

A ray of light originating at O is aimed toward point A on a half-silvered mirror. Half the light is reflected toward mirror one at C and half of it passes through to mirror two at B. The ray at C is reflected back toward A and half of it is transmitted toward D. The ray at B is also reflected back toward A with half of it reflected toward D. The two rays have paths in common between O and A, and between A and D. The parts of the rays that do not have paths in common make round trips in perpendicular directions. The two rays produce interference fringes at D, and it is this fringe pattern that is observed. Since it is a very sensitive measure of wave length and since wave length alters proportionately with the speed of light, a change in the interference pattern was expected as the entire experimental setup (mounted on a massive stone disc floating on mercury) was rotated about a vertical axis. However, while a large number of experiments with the apparatus occupying many different orientations with respect to the fixed stars was carried out, the displacement of the interference fringes always remained well within the errors of observations. It was impossible to detect any variation in the speed of light. [Figure 53]

Figure 53

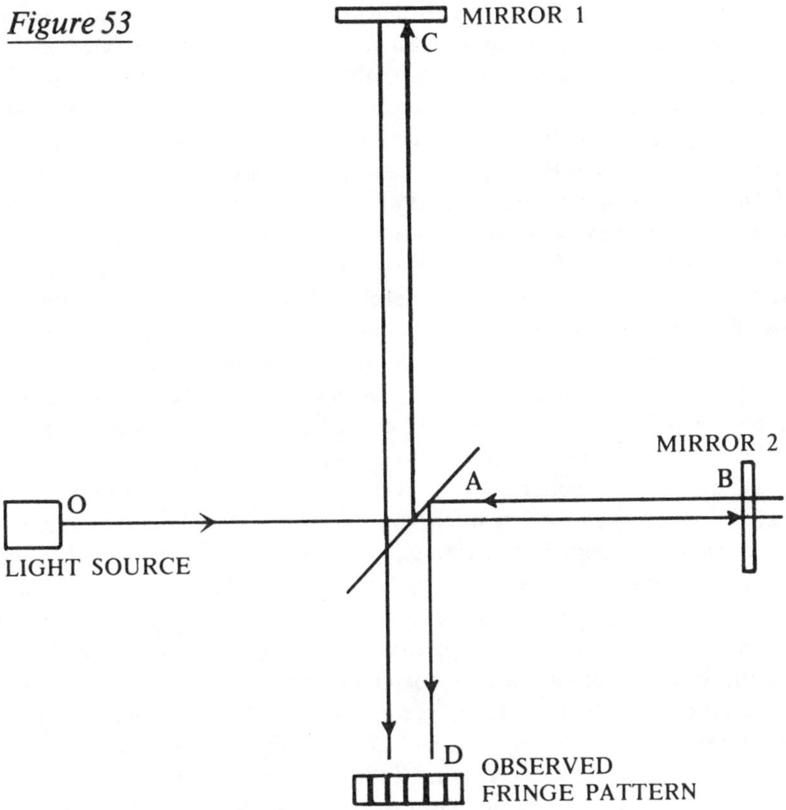

The Michelson-Morley Experiment

Stokes's variant of Fresnel's theory, however, left room for explanations which were little more than ad hoc restatements of the results. They were not in accord with the principle of sufficient reason. Furthermore, in 1892, Oliver Lodge designed an experiment which showed that the velocity of light is not affected by the motion of adjacent matter to the extent of 1/200 parts (a half-percent) of the velocity of the matter. He bifurcated a beam of light which was made to travel in opposite directions round a closed path in space between two rapidly rotating steel disks.

It was Lorentz who developed a relativity theory which did not depend upon ad hoc assumptions about the behavior of a ponderable aether, but rather upon the immanent geometry of the electromagnetic field.

Lorentz developed, along with his force law, the more general Lorentz transformations which mathematically correct for the apparent constancy of the speed of light. [Figure 54] He postulated that the aether is always at rest, in effect, dispensing with it for purposes of his theory as Einstein pointed out. He accounted for the failure of the interference pattern to change, despite rotation of the mirrors in Michelson and Morley's experiments, by the relativistic implications of his force law. Lorentz extrapolated from the force law $\mathbf{F}=q(\mathbf{E}+\mathbf{v}\times\mathbf{B})$ to produce the transformations in the electromagnetic force field which occur when a system in motion is viewed alternately by an observer at rest relative to the motion of the system, and an observer traveling at the same velocity as the system who therefore considers the system to be at rest. If we suppose the existence in the system of both an electric force field **E** and a magnetic force field **B**, then, in addition to noting the field strength **E**, the moving observer will note an additional force due to the motion of the magnetic field giving the transformation $\mathbf{B}'=\mathbf{B}-(\mathbf{v}/c\times\mathbf{E})$. Similarly, the moving observer will detect a magnetic field caused by the motion of the electric field $\mathbf{E}'=\mathbf{E}+\mathbf{v}/c\times\mathbf{B}$. Therefore, when **E** and **B** are equal to zero, $\mathbf{E}'$ and $\mathbf{B}'$ will also equal zero. Dark areas within the interference pattern, areas of zero magnetic and electric intensity, will remain dark when the velocity of the system is changed (by rotating the apparatus).

The shape of the interference pattern will seem to have changed from circular to elliptic to an observer who does not share in the changed motion, but this will not be apparent to an observer moving with the system, who will compensate for the change in velocity by automatically adjusting his time and distance scale. Lorentz made the mathematical adjustment to slow down time and shrink length in the direction of motion by the factor $\sqrt{1-(v^2/c^2)}$. In this way, the failure to register the difference in time between the two trips in the Michelson-Morley experiments is compensated for. He did not offer a satisfactory explanation for the adjustment. Einstein drew the general axiomatic conclusion that space and time are calculated according to criteria of simultaneity, but that simultaneity is actually relative to motion. The argument is simple enough. Light sent out in two directions from the middle car of a moving train will reach the ends simultaneously, according to the perception of a viewer on the train who is unaware of his motion. To an outside observer, this will not be the case. The rest follows.

Wave Interference Patterns

In order to better understand the Michelson-Morley experiments and the full implications of the wave nature of electromagnetic radiation, we will follow the work of Young and Fresnel.

In 1801, Young proposed a wave theory of light:

Figure 54

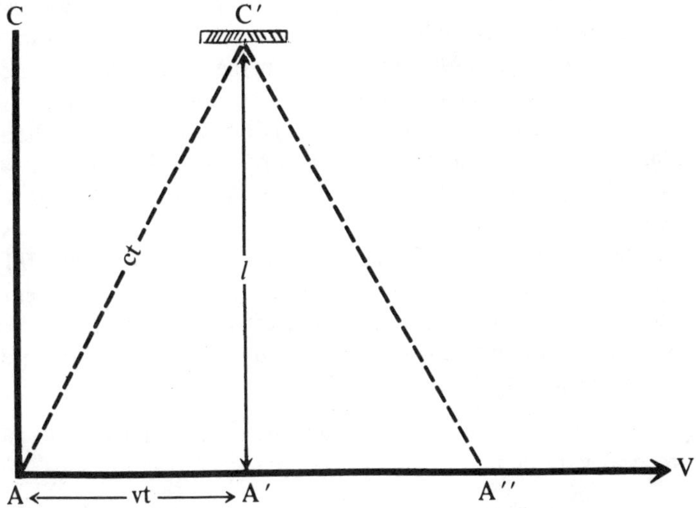

Time of travel along the horizontal and vertical arms of the Michelson-Morley apparatus. While light is passing to mirror C the apparatus is moving to the right at velocity v. Hence, the light actually arrives at the mirror when it is at C′. During the time t that light travels the distance ct from A to C′, the apparatus travels the distance vt from A to A′. From the Pythagorean theorem, $(ct)^2 = l^2 + (vt)^2$, or

$$t = \frac{l}{c}\frac{1}{\sqrt{1-v^2/c^2}}.$$

The time for light to travel to mirror C and return to A along the path AC′A″ is twice this time, or

$$T_C = 2t = \frac{2l}{c}\frac{1}{\sqrt{1-v^2/c^2}}.$$

A return trip to mirror B shown in Figure 53 can be accomplished in a time

$$T_B = \frac{l}{c+v} + \frac{l}{c-v} = \frac{2l}{c}\frac{1}{1-v^2/c^2}$$

The difference between these two travel times should determine a phase difference in the light waves when they recombine at the telescope. No such difference has ever been measured in any of the variations performed on this experiment. The Lorentz transformation, which leads to a length contraction of $\sqrt{1-v^2/c^2}$ in the B direction, has been introduced, superseding the Galilean transformation, in light of these results.

> Suppose a number of equal waves of water to move upon the surface a stagnant lake, with a certain constant velocity, and to enter a narrow channel leading out of the lake; suppose then another similar cause to have excited another equal series of waves, which arrive at the same channel, with the same velocity, and at the same time with the first. Neither series of waves will destroy the other, but their effects will be combined; if they enter the channel in such a manner that the elevations of one series coincide with those of the other, they must together produce a series of joint elevations; but if the elevations of one series are so situated as to correspond to the depressions of the other, they must exactly fill up those depressions and the surface of the water must remain smooth. Now I maintain that similar effects take place whenever two portions of light are thus mixed; and this I call the general law of the interference of light.

He was led to thus revive Huygens's wave theory because of experimental work being done in electricity:

> That a medium resembling in many properties that which has been denominated aether does really exist is undeniably proved by the phenomena of electricity. The rapid transmission of the electric shock shows that the electric medium is possessed of an elasticity as great as is necessary to be supposed for the propagation of light. Whether the electric aether is to be considered the same with the luminous aether, if such a fluid exists, may perhaps at some future time be discovered by experiment: hitherto I have not been able to observe that the refractive power of a fluid undergoes any change by electricity.

The major stumbling block to the acceptance of the wave theory of light lay in the original hypothesis that light waves were longitudinal rather than transverse in form. If this were the case, then lightwaves, like sound waves, would be formed by a process of linear condensations and rarefactions of the transmitting medium in the direction of travel. However, as refraction phenomena show, light is a planar (two-directional) phenomenon. Huygens, in the seventeenth century, supported his wave theory by the behavior of two beams of light which, when they intersect, proceed to cross each other — each as if the other did not exist. The same phenomena can be observed with water waves. The corpuscular theory of light would seem to predict that the particles would collide and then scatter. Nonetheless, the double refraction of a ray of monochromatic light penetrating into a crystal, such as an iceland spar, presented a stumbling block to the theory. It will be split into two different rays of the same color yet having different properties. [Figure 55] They do not follow the same path, nor are they refracted in the same way when they emerge from the crystal.

It was Fresnel who solved the problem by showing that light waves were, like the waves on the surface of the ocean, transverse. [Figure 56] In sound waves, the condensation and rarefaction occurs in the same direction of the propagation of the wave impulse. With light, the electromagnetic impulse varies along a wave front that is perpendicular to its direction of propagation. Light passing through a crystal can be screened through the crystal's structure in two different directions. (A longitudinal wave is determined in its direction of variation of impulse; a transverse wave only loses one degree of freedom by the constraint that it vary in a direction perpendicular to the line of propagation.)

An important objection raised against the wave theory was the sharp ray-like character of light shown, for example, by the sharp shadows cast by light, as compared to the behavior of sound which does pass around corners. Fresnel demonstrated that if light were made to pass through a tiny hole whose dimensions were of the order of a wave length, this sharpness would disappear. In comparison sound waves are huge compared to the dimensions of a screen placed between a source and an observer, for example. When sound is made to pass through a slit whose width is far greater than the wave length of the waves, the same phenomena occurs. This experiment is realized when an explosion occurs in a mountain gorge leading to an open plain.

Young was able to produce regular interference patterns by projecting a coherent beam of light through two closely placed slits. For the purpose, he used sodium salt ignited in a bunsen burner. Modern lasers do nicely as a substitute. In practice, even light of the same color is not coherent. It emits regular waves only over extremely short periods of time, in the order of 10^{-3} seconds. This is a function of the same sort of structuring of quantum-type effect which we discussed in connection with the photoelectric effect. Owing to the random nature of the change in phase of light, we cannot expect the waves emitted from two distinct monochromatic point sources to be in phase.

Whenever we wish to obtain interference effects from the superposition of two monochromatic beams, we must operate with beams that are coherent. Suppose that we had two point sources S_1 and S_2 emitting the same monochromatic light. If at a point P and at the instant t the waves issuing from the two sources are opposite in phase, they will tend to cancel each other at this instant. [Figure 57] The cancellation will be complete if the amplitudes and hence the intensities of the two sets of waves are the same. But, for this cancellation to betray itself to the eye by an absence of luminosity, it must endure over a certain period of time. This coherence of

Figure 55

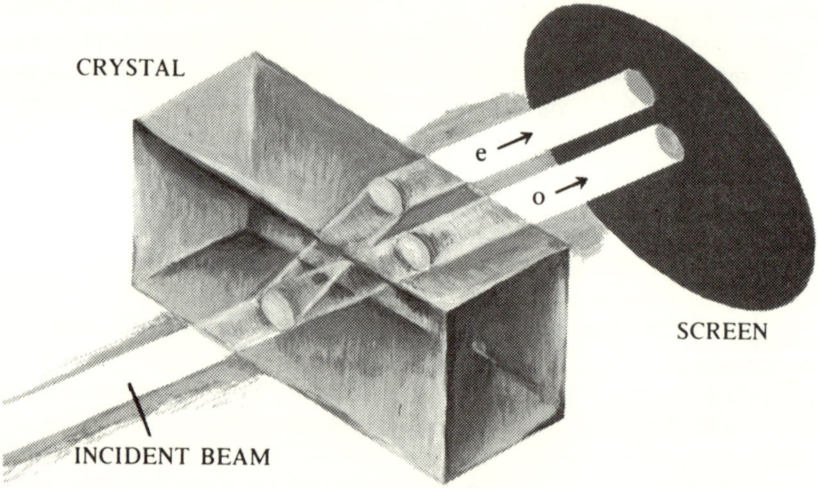

A beam of unpolarized light falling on a birefringent crystal, such as Iceland spar, is split into two beams which are polarized at right angles to each other.

---

phase is one of the important properties of a laser which is associated with concentration of focus.

Young's experiment was the first to furnish concrete evidence of the wave nature of light. He took a screen with two fine holes, S_1 and S_2, placed near to each other. These were placed before a point source S of monochromatic light. If the two holes are sufficiently close, the light will be divided into two separate beams which will overlap. They will interfere with each other as they travel through space to produce an interference fringe on a sheet of paper placed on the other side of the screen.

Consider any point P on the sheet of paper. [Figure 58] It receives light from both holes. The wave motion at P results from the superposition of the two sets of waves which reach P from S after having passed through the holes S_1 and S_2 respectively. Since a superposition of two harmonic vibrations of the same frequency yields a harmonic vibration having this frequency, we are certain that the vibratory disturbance at P will be harmonic. The only question to be settled is the amplitude of this vibratory motion. Suppose, then, that the distances S_1P and S_2P differ by a half wave length of the light or by any odd multiple of a half wave length. Waves from S_1 and waves from S_2 will then be in opposite phase when they reach P. A crest and a trough will exactly counterbalance each other to create a dark fringe. On the other hand, if the distances S_1P and S_2P do not differ or differ by a multiple of the wave length, then the two sets of waves will have the same phase at the point P. Reinforcement will occur and we see a bright fringe.

Electromagnetic Radiation

Heinrich Hertz was the first person to verify that electromagnetic radiation was propagated at the speed of light. He did this by establishing the existence of electromagnetic waves, their length and frequency. He was able to do this by the expedient device of reflecting them, thereby creating standing waves. [Figure 59] The experiment can be conveniently replicated by using a ham radio as a transmitter for the waves, and a small electric light bulb attached to a coil as a receiver. The waves will be within range of laboratory demonstration. AM radio waves are in the range of 100 million cycles per second. Their length is approximately three feet.

Hertz performed a series of experiments beginning in 1886, when he noticed an effect which formed the starting-point of his later researches. When an open circuit was formed of a piece of copper wire, bent into the form of a rectangle, so that the ends of the wire were separated only by a short air-gap, and when this open circuit was connected by a wire with any point of a circuit through which the spark discharge of an induction coil was taking place, it was found that a spark passed the air gap of the open circuit. (Although we have not discussed sparking at length, the reader should be familiar with the effect which occurs when oppositely charged bodies are placed in close proximity, but not touching. A spark will jump the gap to equalize the charge. The production of such a spark is not simply a matter of electron flow, but involves interaction with the medium between the charges.)

Hertz explained this by supposing that the change of potential which is propagated along the connecting wire from the induction coil reaches one end of the open circuit before it reaches the other so that a spark passes between them. This phenomenon indicated that electric potential was propagated at a finite speed. Hertz then found that a spark could be

Figure 56

Certain basic information about waves is summarized here.

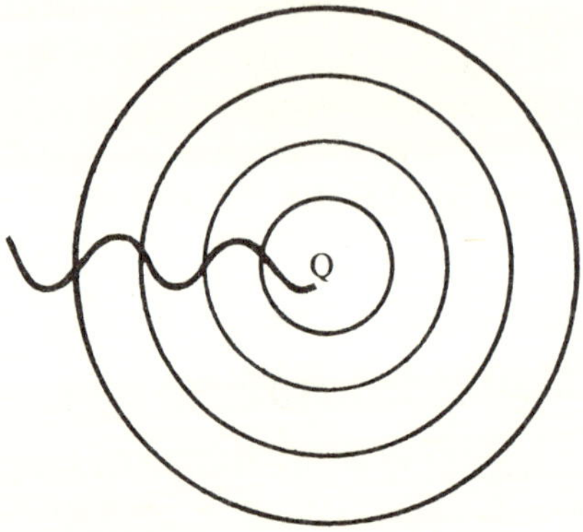

A. For a spherical wave emitted from Q, maxima, minima, and, more generally, points of equal phase are located on spheres concentric on Q. This kind of wave can be produced in two dimensions by dropping a pebble into water.

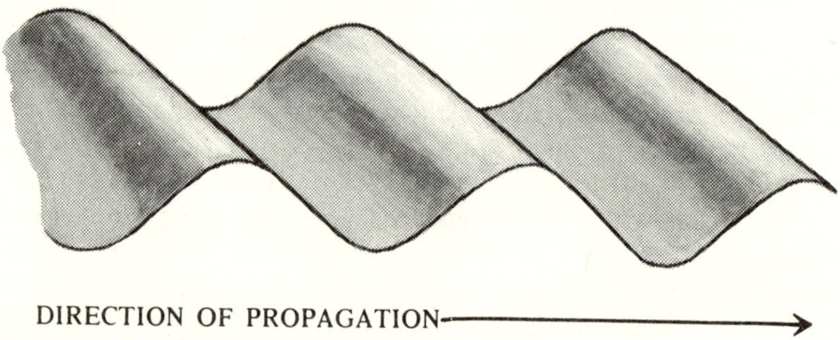

B. For a plane wave, phases are equal on planes perpendicular to the direction of propagation.

---

induced in the open or secondary circuit even when it was not in metallic connection with the primary circuit in which the electric oscillations were generated. This was possible despite the weakness of the primary discharge because the secondary circuit was of such dimensions as to make the free period of electric oscillations in it nearly equal to the period of the oscillations in the primary circuit. The disturbance which passed from one circuit to the other by induction was greatly intensified in the secondary circuit by the effect of resonance.

In 1888, Hertz showed that electromagnetic action is propagated in air at a finite velocity. He transmitted the disturbance from the primary oscillator by two different paths, through air and along a wire. In this way,

Figure 56 Cont.

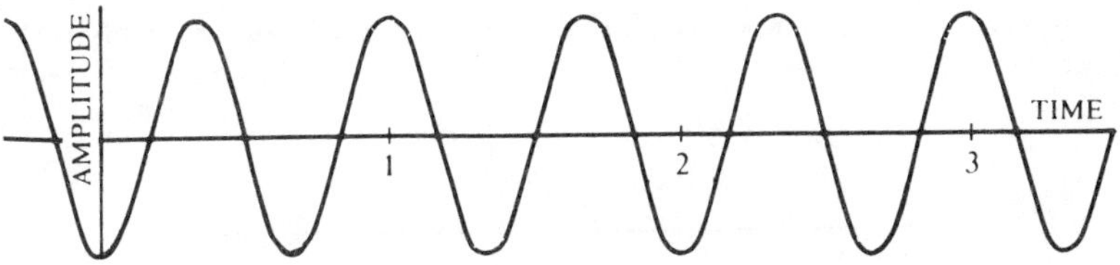

C. The wave frequency is the number of completed cycles per unit time, here 1.5 cycles per second. The amplitude is the maximum displacement of the wave from the null line. The wavelength is the spatial displacement between crests in a wave train. The product of wavelength and frequency is the wave propagation velocity.

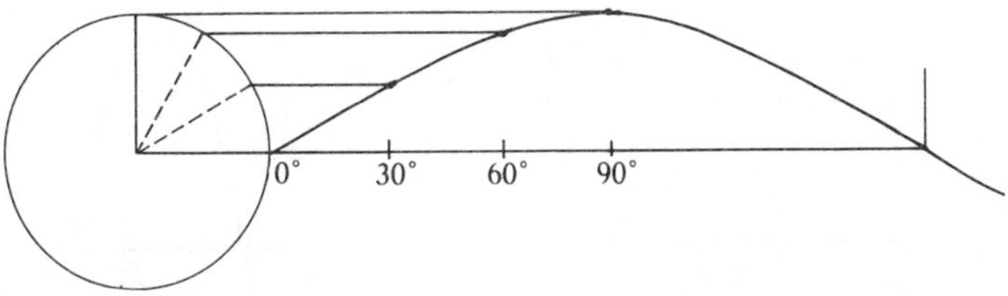

D. The production of the sine function graphically from the representation of the trigonometric functions on a unit circle. On the right triangle formed with the radius of the unit circle as the hypotenuse, the vertical arm is a representation of the magnitude of the sine function of the angle at the center of the circle. The horizontal arm represents the cosine function for the same angle. The sine and cosine functions are the most regular types of waves. Fourier proved that any function can be represented as a summation of these functions. Using this principle, radio and television transmitters and receivers can encode and decode information that we perceive as sound and light waves.

---

he created interference between them which he could detect with his sparking device. Following this, he showed that electric waves in air are reflected off the surface of a wall. In this way, standing waves may be produced by creating interference between direct and reflected beams.

By showing the interference patterns, he was simultaneously able to demonstrate the wave nature of electromagnetic radiation and attempt to verify the speed of its propagation to show that it was equal to the velocity of propagation of light. He was not successful in the latter effort, although later experimenters were. He had extreme difficulty in controlling wayward radiation which, of course, distorted the effects which he was seeking. His book, *Electric Waves,* is both an interesting account of his

Figure 57

Interference is a temporary amplification or cancellation of superposed waves.

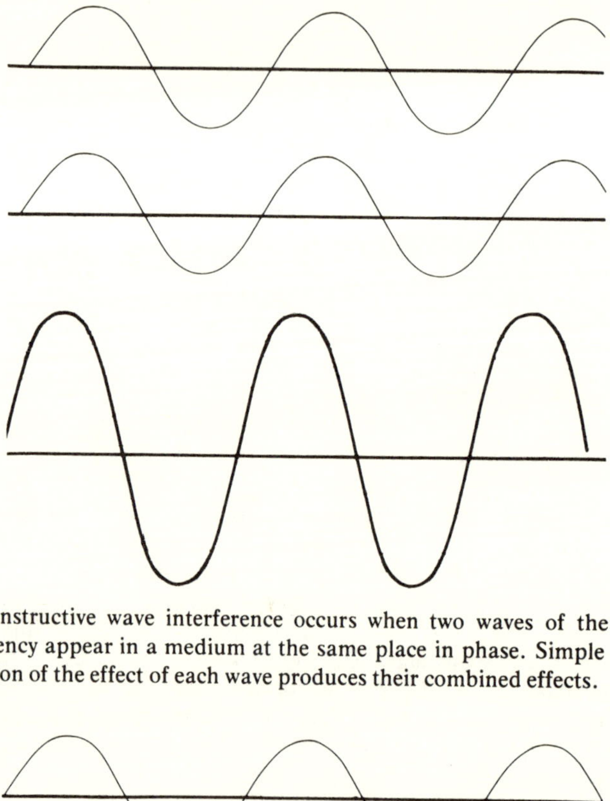

A. Constructive wave interference occurs when two waves of the same frequency appear in a medium at the same place in phase. Simple scalar addition of the effect of each wave produces their combined effects.

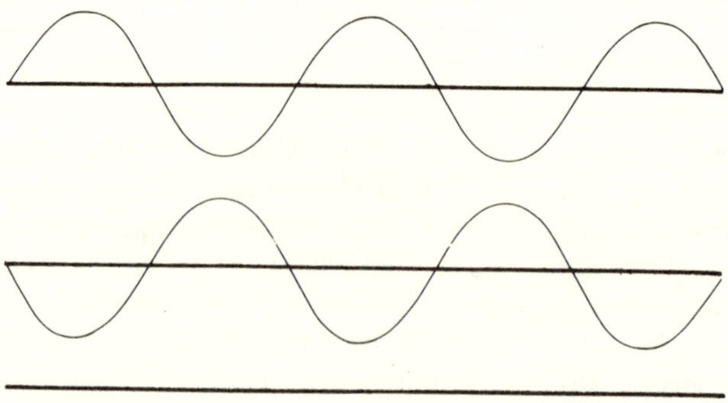

B. Destructive interference occurs when the two waves are half a wavelength out of phase. Then, the crest of one lies on the trough of the other and the effect is complete cancellation when the amplitudes are identical.

Figure 58

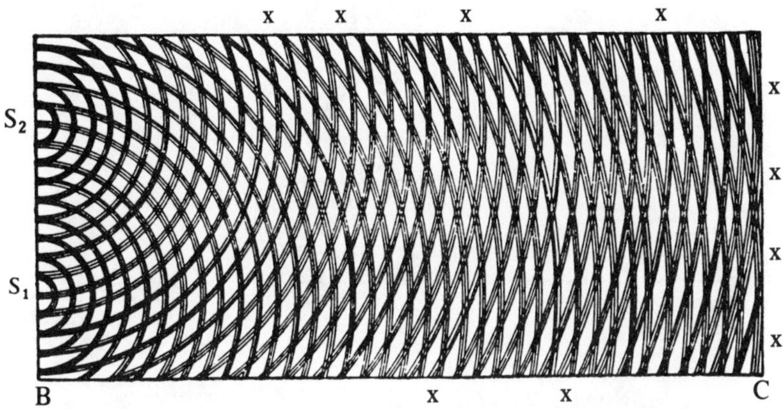

Thomas Young's original drawing shows the interference effect in overlapping waves from two sources S_1 and S_2. By placing the eye near the left edge while viewing the diagram at a grazing angle, the regions of positive interference, marked x, become clearly distinguished.

experiments and a moving description of those moments of intense anguish which he faced when his experiments seemed to demonstrate an infinite velocity of propagation of light. (Due to extraneous radiation, he was, at first, unable to pick up any interference pattern.)

While Hertz tested for the finite velocity of light, he took for granted that radiation was propagated as a wave through a continuous medium. It is true that no liquid is microscopically continuous, but that is not the point in question. All of the aether theorists of the nineteenth century assumed that the electromagnetic field could be treated as a continuous fluid with zero divergence. In this, they were proven wrong by the discovery of the photoelectric effect and photons. Nonetheless, they succeeded in establishing those wave-like properties of radiation which we have just studied.

To bridge the gap between wave and particle geometries of radiation, it will be necessary to define the essential singularities which render possible a geometry which subsumes them. Such a higher-order geometry will not develop merely as a theoretical enterprise. It will necessarily emerge as a byproduct of work now in progress on nonlinear effects in plasmas, to subsume those only apparently linear effects known to those scientists of the nineteenth century who studied the electromagnetic field.

Figure 59

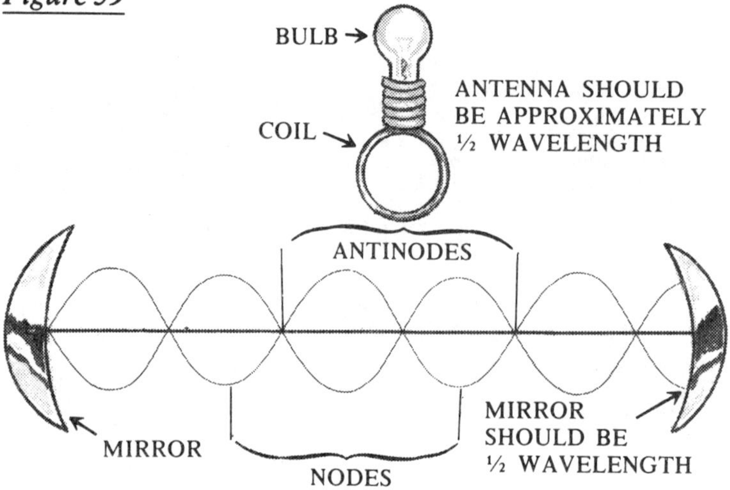

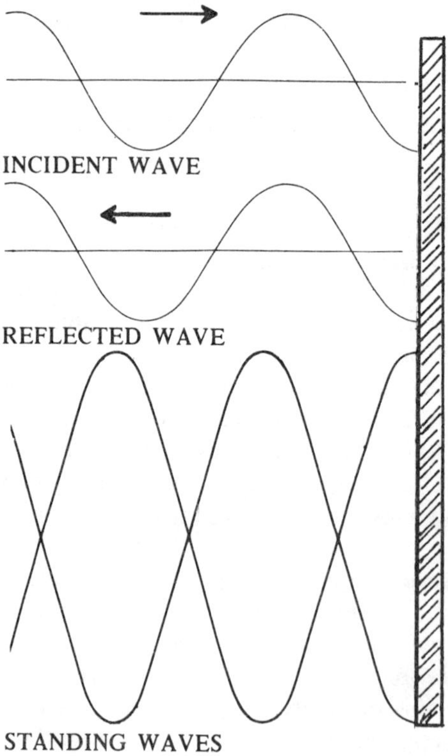

In the Hertz experiment, standing waves are produced by reflecting radio waves from a parabolic antenna "mirror." The characteristic standing wave pattern — nodes and antinodes alternately at intervals of one quarter wavelength — is observed using a coil of wire attached to a flashlight bulb. The bulb lights at the antinodes, but is dark at the nodes. The electromagnetic field at the antinodes generates a current in the coil which heats the bulb filament, just as in the case of an ordinary light bulb. For best results, the coil should be about one-half wavelength.

Relativistic Truth

It is typical to scorn "aether theories" with the blithe assertion that Einstein "proved" that they are superfluous. In his later years, Einstein himself asserted that the had only sought to bury the notion of a ponderable aether. He continued to search for a replacement in his quest for a unified field theory. The statement by positivists that the electromagnetic field is merely a geometric entity is worse than a subterfuge. The field is geometric precisely because it is physically real. As Riemann proved, there is only physical geometry. The abstraction of geometry from physics only leads to blunders and sterile formalisms.

There is no ponderable aether. The medium from which matter is created, in which it perpetuates itself, and through which it transmits electromagnetic and gravitational effects is not a collection of material particles. It is other than ponderable. There is a physical particle-field collectivity.

Descartes's vortex theory, despite Newton's trivial criticisms of its formal characteristics, is, in the final analysis, the only acceptable basis for a unified field theory. Matter can only be created by a self-reproductive process within the field itself which, having emerged from the field, in turn redefines it.

Descartes was correct in postulating a vortex theory. Newton was necessarily wrong. A particle theory, by definition, cannot explain the existence of either particles or fields.

The distinction is that between relative and relativistic truth. Cultural relativists use banalizations of relativity theory to assert that since motion is relative and the calculations of space and time intervals are, therefore, not absolute, then, likewise, there is no absolute criteria for truth. The deer calls the hunter a bad man, while the hunter finds the deer to be a good meal... and so on, endlessly. Such theorizing is not merely naively wrong. Bertrand Russell and his gang deliberately distorted science in an effort to discredit morality. Their aim was to say that there is no absolute criteria for morality at any given time, since from one period to the next such criteria may be superseded.

To refute this, it is only necessary to refer back to the case of Descartes. It is the Cartesian tradition of science, not Newton's *Principles of Natural Science,* which was necessary to the later work upon which the discovery of electromagnetic field theory depended. Newton corrected certain errors by Descartes; to that extent, he advanced our understanding of the universe. He used these minor corrections to justify his reductionist epistemology. To the extent that British empiricism became hegemonic thereby, he set back scientific progress by damaging the possibility for creative thinking in future scientists.

By relativistic truth, we assert that *yes* truth changes. What seems true in one period is recognized as merely an approximation by later scientists. The depth of our penetration of the universe continually increases. And, in the act of discovery, we change the universe. It is we, the highest expression of life, who have molded the biosphere. Tomorrow, we will stretch out further into the universe at large. Not only will our understanding of the laws of the universe change, but we will use that understanding to willfully change those laws.

We will know truth and morality — relativistic truth. The criteria which will guide us will be the same commitment to that scientific progress which alone will allow our human species to survive. This is our commitment to the negentropic process of continual evolution of the laws of the universe from which we, the human species, have emerged. This is the truth of relativistic truth.

A Suggested Sequence of Experiments for Demonstration Or Laboratory Classes

The Hertz experiment is an excellent point of departure. It lays out the way that "space" can be structured by the production of radiation. The velocity of light can be computed by computation of the wave length and frequency of waves. Interference can be studied. The student should attempt to contrast the mental picture of an electron theory and the alternate Maxwellian conception of a mediating magnetic field in attempting to describe the production of waves.

The Young-Fresnel experiments should be repeated, using a laser to guarantee coherent light. The student's mental picture should now be extended to subsume both light and radio waves. Experiments with polarizing screens can be performed easily to show the spatial dimensionality of light. The relationship of the screens to each other determines the degree of screening of light. If possible, students should also work with a double polarizing substance like iceland spar.

The generation of electricity should be studied next. Attention should be paid to the how changes in the primary-secondary relation between the wire conductors effects the current generated. The position of the two coils should be varied. The amount of loops in each should be varied. The direction in which the current is introduced in the first should be changed. Then the pattern of alternation in the first and generation in the second should be examined. Finally, the coils should be moved together and apart. The last experiment should be replicated with a magnet.

The magnetic effects of current should now be studied by the construction of coils of wire — electromagnets. Again, the direction of the current should be alternated. Faraday's "lines of force" may be shown by iron filing patterns. Students may wish to test the screw field analogy.

Electrostatic induction and conduction effects such as Franklin's experiments on electricity are easily reproduced. The photoelectric effect and similar such experiments are available in motion pictures prepared for classroom use and in general circulation.

Students should use the Hertz experiment as a standard by which to gauge the other demonstrations, returning to it in the end with the purpose of determining for themselves how far we have penetrated to the structuring of space and, conversely, the open questions left to us to solve. Aether theories were not trivial, mechanistic notions, even though they were wrong. The question still remains before us: What waves?

It would be most desirable if the student could also take apart a radio set and a galvanometer. The most sophisticated measuring instrument is an oscilloscope. It is possible to build a simple one. Students should become familiar with the workings of both the simple models and the highly sensitive ones in use in laboratories. While motors are extremely familiar, the advanced techniques of electrical generation through nuclear fission and fusion power should be studied. The newest frontier of electrical generation is directly through plasma-magnetic field interaction.

Bibliographical Notes

Blank, A.A.; Friedrichs, K.O.; and Grad, H. *Notes on Magneto-Hydrodynamics, A Theory of Maxwell's Equations Without Displacement Current.* New York: Courant Institute of Mathematical Sciences, 1957.

Bostick, Winston H. "The Pinch Effect Revisited." *International Journal of Fusion Energy,* 1:1-48.
This article should be read through repeatedly. It locates exactly those non-linear effects which reveal themselves in the frontiers of present-day plasma research. The article further benefits from a genuine popular style which neither condescends nor deliberately mystifies the reader.

Cohen, I. Bernard, ed. *Benjamin Franklin's Experiments, A New Edition of Franklin's Experiments and Observations on Electricity.* Cambridge, Mass.: Harvard University Press, 1941.

Courant, Richard. *Differential and Integral Calculus.* Translated by E.J. McShane. New York: Interscience Publishers, 1937.
This book is a necessary antidote to falsely rigorous modern textbooks which locate rigor as an abstract algebraic exercise. This is a rigorous text which treats the calculus from the standpoint developed by Leibniz.

Einstein, A.; Lorentz, H.A.; Minkowski, H.; and Weyl, H. *The Principle of Relativity.* New York: Dover, 1923.

Graf, Rudolf F. *Safe and Simple Electrical Experiments.* New York: Dover Publications, 1973.

Hertz, Heinrich. *Electric Waves.* New York: Dover, 1962.

Kellogg, Oliver Dimon. *Foundations of Potential Theory.* New York: Dover, 1954.
This is an excellent extended treatment of the subject which includes, in particular, harmonic functions.

Kline, Morris. *Calculus: Intuitive and Physical Approach.* New York: Wiley, 1967.
Both volumes should be read in conjunction with the Courant text for reasons of contrast, since Kline is a "Newtonian," but also because Kline connects the mathematical development directly to problems from mechanics.

Knopp, Konrad. *Elements of the Theory of Functions.* Translated by Frederick Bagemihl. New York: Dover, 1953.
An excellent introduction to the study of the geometry of complex number functions.

Lorentz, H.A. *Theory of Electrons and Its Applications to the Phenomena of Light and Radiant Heat.* New York: Dover, 1952.

Maxwell, James Clerk. *A Treatise of Electricity and Magnetism.* 2 vols. New York: Dover, 1954.

Parpart, Uwe. "The Concept of the Transfinite." *The Campaigner,* Vol. 9, Nos. 1-2.

This is an essential companion work to this book. It contains an analysis and criticism of the history of nineteenth century mathematics from the standpoint developed in this book, including an extended analysis of the failure of Einstein to understand Riemann's theory of relativity. It also includes the only English translation available of Georg Cantor's 1883 *Grundlagen.*

Parpart, U.; Bardwell, S.; and Levitt, M. *Nonlinear Phenomena in the Physical Sciences.* New York: University Editions, forthcoming.

The article by Bardwell, which first appeared serially in the *Fusion Energy Foundation Newsletter,* is a comprehensive survey of nonlinear effects already discovered by plasma scientists, and was both an inspiration for and a confirmation of the thesis developed in this book.

Prandtl, L., and Tietjens, O.G. *Fundamentals of Hydro and Aeromechanics.* New York: Dover, 1957.

This book is an excellent reference to the notions of fluid dynamics which underlie nineteenth century aether theory. There is an especially valuable treatment of the work of Helmholtz and Thompson in developing a theory of vortices which ironically uses the behavior of the mechanical field according to the Faraday model as an intuition for the study of real fluids.

Priestly, Joseph. *The History and Present State of Electricity, With Original Experiments.* Vol. I and II. New York: Johnson Reprint Corp., 1966.

Scott, William Taussig. *The Physics of Electricity and Magnetism.* New York: Wiley, 1966.

This book has all the cited failings of modern texts; however, it does make more connections to actual experimental practice than others which I have seen. The reader is advised to survey available texts and make his own selection.

Williams, L. Pearce. *Michael Faraday a Biography.* New York: Simon and Schuster, 1965.

The reader is strongly advised to read this as a counterposition to the point of view developed in the present work. The author is partisan to Faraday's world view. The book is also useful for its detailed treatment of the experimental work of Faraday and others.

Whittaker, Sir Edmund. *A History of the Theories of Aether and Electricity.* New York: Harper, 1960.

This book, published in two volumes, is an invaluable history of the subject and was an unfailing source for the present work.

Translator's Preface

Georg Friedrich Bernhard Riemann, the eminent nineteenth century mathematical physicist, was born September 17, 1826 in Breselenz, a village in the Kingdom of Hannover, Germany. His education, upon the completion of his studies at the Johanneum in Lüneberg, included study under the great German mathematician Gauss at the University of Göttingen (1846-1847), and Dirichlet and Jacobi at the University of Berlin (1847-1849). He returned to the University of Göttingen in 1849 where he completed his dissertation in 1851.

Still the most complete biography extant on Riemann is the one written by his colleague Richard Dedekind at the request of Heinrich Weber, and based on memory and Riemann's letters to his family. All of the various anecdotes concerning Riemann which have been passed down through generations in the equally numerous histories of mathematics are based on the brief Dedekind biography. But, these histories fail to make mention of what Riemann considered the stifling and oppressive intellectual atmosphere which surrounded the University of Göttingen. As Dedekind notes, when Riemann's deteriorating health forced him to take extensive trips to Italy, "there he felt himself to be freer than most men, without the inhibiting considerations which he believed that he had to take in stride in Göttingen."

Such terminology, which Dedekind obviously carefully chose, obscures and personalizes what was an international effort to contain Riemann and the Franklinesque scientific networks that were making critical advances in mathematics and physics during the nineteenth century. The best documented evidence of the effect these "inhibiting factors" had on Riemann's own development and his ability to influence other scientists through his works is his poor publishing record. Of the 36 papers and reports published in his *Collected Works* and *Supplement*, Riemann had only published 11.

Dedekind attributes this small number of published papers to Riemann's painstaking insistence on perfection, but a footnote by Heinrich Weber at the conclusion of Riemann's "A Contribution to Electrodynamics" (included in this book) indicates that quite another process was going on. This paper was retracted allegedly because

Riemann found an error in it. On the contrary, there is no mistake in it at all, as White shows in the preceding section of this book. It is highly doubtful that Riemann would have withdrawn his paper because he held the same mistaken opinion of it that Clausius did after the work's posthumous publication — an opinion quoted by Maxwell in his dismissal of Riemann's "Contribution to Electrodynamics."

From 1851 until 1858, according to the Dedekind biography, the only trips Riemann made outside of Göttingen were recuperative visits to the nearby Harz Mountains. But, beginning in 1858, the allegedly shy Riemann began a series of journeys to meet with leading European scientific circles that can only be characterized as his way of breaking the restrictive atmosphere of Göttingen and of publicizing his ideas far beyond the circulation of the small number of his published works. In 1858, Riemann initiated a long friendship with three leading Italian mathematicians — Brioschi, Betti, and Casorati — when they stopped in Göttingen while on a tour of Europe. In September 1859, both Riemann and Dedekind traveled to Berlin to meet the leading mathematicians of the Berlin Academy, including Kummer, Borchardt, Kronecker, and Weierstrass. He spent Easter 1860 with the leading mathematicians of the Parisian Academy, Serret, Hermite, Puiseux, Briot, and Bouquet.

Riemann continued to seek out the key scientific networks during his stays in Italy from 1862 until his death in Italy in 1866, visiting Palermo, Naples, Rome, Livorno, Pisa, Florence, and Milan, not only for their art treasures, but also for the rich collaborative environment that nourished scientific research in these cities. In Pisa, he became close friends with Betti, Felici, Novi, Villari, Tassinari, and Beltrami. The latter even offered him a professorship at Pisa in 1863, which Riemann was forced to decline because of his health. Although he traveled back to Germany to teach some classes at Göttingen, he considered his time in Italy to be the happiest years of his life. He died in 1866 and is buried in Biganzolo.

It is yet unclear what were the precise inhibiting considerations impinging on Riemann at Göttingen, not so in the case of Riemann's intellectual successor Georg Cantor. As Uwe Parpart shows through Cantor's own letters, Berlin mathematics czar Leopold Kronecker, who was also a contemporary of Riemann, created such hostility toward Cantor's work in French and German mathematics circles that Cantor was pushed to a complete mental breakdown and subsequent passive acceptance of Kronecker's ridicule of his groundbreaking work on the transfinite. (See Parpart, Uwe. "The Concept of the Transfinite," *The Campaigner*, Vol. 9, Nos. 1-2.)

The twentieth century is not devoid of its Kroneckers either. The late Werner Heisenberg is exemplary in his denial of progress through ever higher levels of scientific development. The following is excerpted from his book *Across the Frontiers* (Harper and Row, 1974):

> In physics, Einstein's achievements were in the highest degree revolutionary, their consequences reaching out far beyond the science to which they initially belong. Yet, however paradoxical this may sound, in important aspects of his nature Einstein's was a conservative mind. Through his years of development, he had become wedded to the nineteenth century belief in progress.

Against this nominalists' commitment to zero growth are those mathematicians, physicists, and biologists who are, with intellectual vigor, confronting problems at the very frontiers of science. To these men of science, a belief in progress is at the heart of scientific advancement.

SOURCES

Bernhard Riemann's physics text, *Gravity, Electricity and Magnetism* (*Schwere, Elektricität und Magnetismus*), was compiled from lecture notes by Karl Hattendorf, a particularly trusted student of Riemann's who, in 1865, was delegated the task of completing Riemann's work on minimal surfaces. The physics text was published posthumously in 1876 by Carl Rümpler in Hannover, Germany.

As Hattendorf states in his introduction to the lectures, he alone is responsible for the text and explication, while the method and included calculations are Riemann's. In this first English translation from the German, I have excerpted sections 54 through 104. The earlier deleted sections contain Riemann's discussion of material on introductory physics fully elaborated in this and other texts on the subject, while the deleted sections 105 through 111 contain the discussion on earth magnetism which is largely outdated.

Riemann's physics text is one of four books compiled from lecture notes and posthumously published during the last quarter of the nineteenth century. His *Elliptical Functions* (*Elliptische Functionen*) was published by B.G. Teubner with additional material by H. Stahl in Leipzig in 1899. *Partial Differential Equations in Mathematical Physics from Riemann's Lectures* (*Die partiellen Differential-Gleichungen der mathematischen Physik nach Riemann's Vorlesungen*) was compiled by H. Weber and published by F. Vieweg and Son in Braunschweig. By 1935, this text had gone through eight printings, making it a standard text of that period. Finally, Riemann's *Lectures on Partial Differential Equations and Their Application to Physical Questions* (*Partielle Differentialgleichungen und deren Anwendung auf physikalische Fragen*

Vorlesungen) was compiled by Karl Hattendorf and published by Brunswick in 1896.

In preparing the translation of the appended paper by Riemann, I have used the *Collected Works* (*Gesammelte Mathematische Werke und Wissenschaftlicher Nachlass*), which was edited by Richard Dedekind and Heinrich Weber, and first published in Leipzig in 1876; and the *Supplement to the Collected Works* (*Gesammelte Mathematische Werke Nachtrage*), edited by M. Noether and W. Wirtinger, and published by B.G. Teubner in Leipzig in 1902. Dover Publications republished this edition and the second 1892 edition of the *Collected Works* in their original German in 1953.

James J. Cleary, Jr.
August 15, 1977

GRAVITY, ELECTRICITY, AND MAGNETISM

according to the lectures of

Bernhard Riemann

compiled by

Karl Hattendorf

and

A CONTRIBUTION TO ELECTRODYNAMICS

by Bernhard Riemann

Preface

This book has resulted from the lectures that Riemann gave on gravity, electricity, and magnetism during the summer semester of 1861 in Göttingen. With the exception of some very brief notes, there is no manuscript extant by Riemann himself on these lectures. Thus, I alone am responsible for this presentation.

The friendly reception that my compilation of Riemann's lectures on partial differential equations has found among all those experts on the subject leaves me hope that this present book will not be unwelcome to friends of Riemann and those studying mathematics.

Just as in partial differential equations, here too we have to thank Lejeune Dirichlet. In addition to his great service to the further development of science it must not be forgotten that it was he who was the first to lecture about partial differential equations and the potential at German universities. These lectures did not end with his death. They now form a regular part of the program at almost all German universities, and Riemann too took over these lectures after Dirichlet. Concerning agreement on the subject matter, then, it is natural that much here agrees with Dirichlet in layout and execution. But Riemann did not limit himself to simply taking possession of his great predecessor's legacy. The connoisseur will discover that he has submitted an abundance of what is characteristically his.

K. Hattendorff.
Aachen, June 24, 1875

FIFTH DIVISION

Galvanic Currents

Section 54
Specific Current Intensity

We will now consider the case in which both kinds of electricity are continuously becoming separated in the conductor. We will presume that the separating forces are known forces.

Every conductor contains an infinite quantity of electric particles. If no effect is to be exerted toward the outside, then the algebraic sum of the quantities of electricity must be equal to zero in every part, no matter how small, of the conductor.

The quantity of electricity ϵ, which is contained in an electric particle, is the measure of the attraction or repulsion which it exerts per unit distance on the electric unit.

We will take an arbitrary surface [Figure 31] in the interior of the conductor and we will consider there to be a surface element ω at any arbitrary place on the surface. The normal of this surface element proceeds

Figure 31

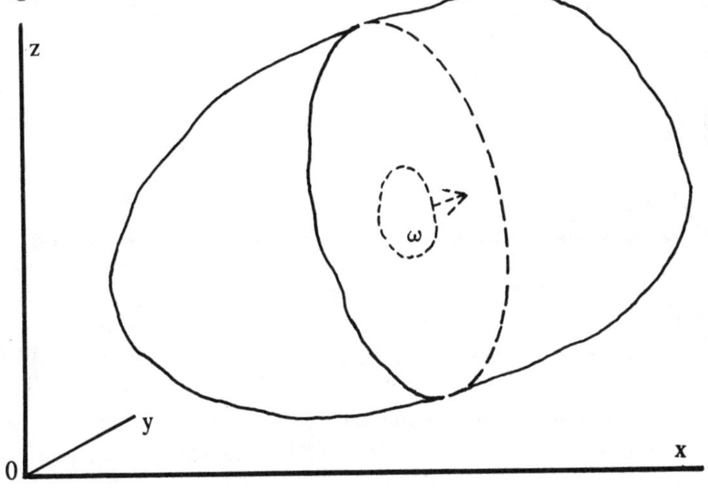

from it in two different directions, which we will differentiate as the direction of the positive or of the negative normal. The *positive* (or the *negative* as the case may be) *side* of the surface faces the space into which the normal enters. During time element dt, very many electric particles go through surface element ω from the negative side of the surface to the positive side, and, conversely, from the positive side over to the negative side.

We will then reduce the algebraic sum of the quantities of electricity which go through surface element ω in time dt from the negative side to the positive side by the algebraic sum of the quantities of electricity which go through the same surface element in the same time from the opposite direction. We will divide the difference by magnitude ω of the surface element and by dt. We will call the quotient the *specific current intensity* in the direction of the positive normal of the surface element. We will denote this by i, then from here on in $i\omega dt$ is the excess of the quantity of electricity which in the time dt flows through the surface element ω from the negative side to the positive side.

Let the surface element lie [Figure 32] perpendicular to the x-axis.

Figure 32

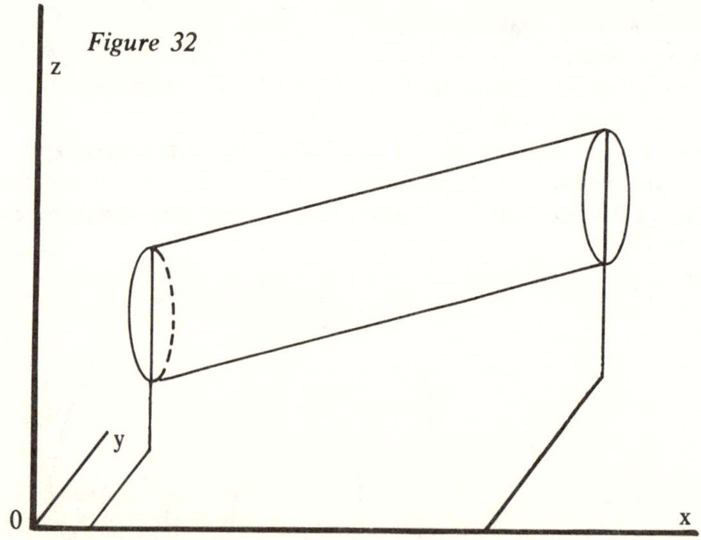

Then, the electric particles will generally go through this element in very different directions, and with very different velocities. To begin with, we will only consider one velocity, v, whose direction forms angles A, B, C with the positive coordinate axes. We will also use this direction as the axis of a cylinder, which has the surface element ω as its base and whose end faces cut out the length vdt on the axis. In this case, v is an *absolute* number. The cylinder's volume is

(1) $\pm\,\omega\, v\, dt \cos A = \pm\,\omega \dfrac{dx}{dt}\, dt,$

since v cosA = dx/dt is the velocity component in the direction of x. In the expression (1) the positive or negative sign holds depending upon whether dx/dt is positive or negative. We will assume that at time t and during the next consecutive time element dt the quantitites of electricity, which are moving with the prescribed velocity v in the given direction, come to be completely and uniformly distributed at an infinite proximity to point (x, y, z). Now, at time t, let the spatial element dS, which touches the point (x, y, z) be filled with electric particles having the velocity in question, and whose quantity of electricity will have the algebraic sum $\Sigma\epsilon$. Then, one can see that

(2) $\pm \dfrac{\sum \varepsilon\, \omega \dfrac{dx}{dt}\, dt}{dS} = \pm \dfrac{\omega\, dt \sum \varepsilon \dfrac{dx}{dt}}{dS}$

is the algebraic sum of the quantity of electricity of those particles which have the prescribed velocity and fill the cylinder at time t + dt. However, it is these and only these particles which have crossed through the cylinder's base ω in the prescribed direction with velocity v during the preceeding time element dt.

If one repeats this for every direction and velocity, then *all* the electric particles will have been taken into account which will have passed through surface element ω in the next successive time element dt after the expiration of time t. One will then get as many expressions of form (2) as there are velocities of different magnitudes and direction. These expressions can be combined into two groups, according to whether dx/dt is positive or negative. For the first group, the expression

(3) $\dfrac{\omega\, dt \sum \varepsilon \dfrac{dx}{dt}}{dS}$

will give the algebraic sum of the quantities of electricity that go from the negative to the positive side of ω during the time element dt which is under consideration. However, one has to extend the summation in (3) over *all* electric particles that occur in spatial element dS of the positive velocity component dx/dt. In a similar manner, for the second group, one obtains

(4) $-\dfrac{\omega\, dt \sum \varepsilon \dfrac{dx}{dt}}{dS}$

for the algebraic sum of the quantities of electricity which cross over from the positive side to the negative side of ω. The summation in (4) is based on all the particles contained in dS, whose velocity component dx/dt is negative. Therefore, in order to calculate the specific current intensity in the direction of x according to the definition, one has to subtract sum (4) from (3) and divide the difference by ωdt. One proceeds in a corresponding manner for the directions of both of the other axes.

Thus, if we denote i_1, i_2, i_3 as the specific current intensities in the direction of the three coordinate axes, then they result in the following equations:

$$(5)\quad i_1\, dS = \sum \varepsilon \frac{dx}{dt},\; i_2\, dS = \sum \varepsilon \frac{dy}{dt},\; i_3\, dS = \sum \varepsilon \frac{dz}{dt}.$$

These summations cover *all* the electric particles that are present at time t in the spatial element dS which is contiguous to point (x, y, z). Every particle's quantity of electricity is to be multiplied by its proper velocity components and all the products thus formed are to be summed.

The specific current intensity can be deduced for any direction from the three magnitudes i_1, i_2, i_3. Let dp/dt be the velocity component of a single electric particle in the direction that forms angles α, β, γ with the positive coordinate axes. Then we have

$$(6)\quad \frac{dp}{dt} = \frac{dx}{dt}\cos\alpha + \frac{dy}{dt}\cos\beta + \frac{dz}{dt}\cos\gamma.$$

If $i_{\alpha\beta\gamma}$ denotes the specific current intensity in this same direction, then what we obtain according to the three equations in (5) is:

$$i_{\alpha\beta\gamma} = \frac{\sum \varepsilon \frac{dp}{dt}}{dS},$$

and what we obtain from this by utilizing equations (6) and (5) is:

$$(7)\quad i_{\alpha\beta\gamma} = i_1 \cos\alpha + i_2 \cos\beta + i_3 \cos\gamma.$$

We will now want to locate a direction in which specific current intensity i is expressed by the equation

$$(8)\quad i = \sqrt{i_1^2 + i_2^2 + i_3^2}.$$

If a, b, c are the angles that this direction forms with the coordinate axes, then they are determined by the equations:

$$(9)\quad \cos a = \frac{i_1}{i},\quad \cos b = \frac{i_2}{i},\quad \cos c = \frac{i_3}{i}.$$

For given these values, equation (7) is transformed into equation (8) if we let $\alpha=a$, $\beta=b$, $\gamma=c$. However, one can also proceed directly from (9) to (8), because, as is well known: $\cos a^2+\cos b^2+\cos c^2=1$.

In order to recognize the significance of specific current intensity i, whose direction has been determined by equations (9), we will introduce the values for i_1, i_2, i_3 of (9) into (7). What results through this is

$$(10)\quad i_{\alpha\beta\gamma} = i\,(\cos a\,\cos\alpha + \cos b\,\cos\beta + \cos c\,\cos\gamma).$$

But the magnitude in parentheses is equal to $\cos\delta$, where δ designates the angle of both directions (a, b, c) and (α, β, γ). Therefore, we have a shorter form,

$$(11)\quad i_{\alpha\beta\gamma} = i\cos\delta.$$

Specifically, we obtain

$$(12)\quad i_{\alpha\beta\gamma} = i \quad \text{for } \delta = 0, \qquad i_{\alpha\beta\gamma} = 0 \quad \text{for } \delta = \frac{\pi}{2}.$$

The direction determined by equations (9) thus has the property that at right angles to it the specific current intensity is equal to zero, while in it, the current intensity is a maximum. Consequently, this direction is the direction of the flow. If one now knows the current intensities in any three directions perpendicular to each other at any point (x, y, z) in the interior of a conductor, then one can find the direction and intensity of the flow according to that same law which is valid for the composition of velocities and the composition of forces. We can call this the parallelepiped law of specific current intensities.

Section 55

Free Electricity. The Equation: $\frac{\partial\varrho}{\partial t}=-\left(\frac{\partial i_1}{\partial x}+\frac{\partial i_2}{\partial y}+\frac{\partial i_3}{\partial z}\right)$

What we mean by *free electricity*, which is contained in a solid element or in a surface element, respectively, is the algebraic sum of the quantities of electricity present. If we divide this sum by the volume of the solid element, or the area of the surface element, respectively, then what results is a quotient which will be called the *density of free electricity* at the place in question. We will denote it as ϱ.

We will now consider an infinitely small parallelepiped in the interior of a conductor, whose edges with lengths of dx, dy, dz will run parallel to the coordinate axes. The corner point lying closest to the origin will have coordinates x, y, z. [Figure 33] The question here is calculating the quantity of electricity by which the free electricity increases inside of the paral-

Figure 33

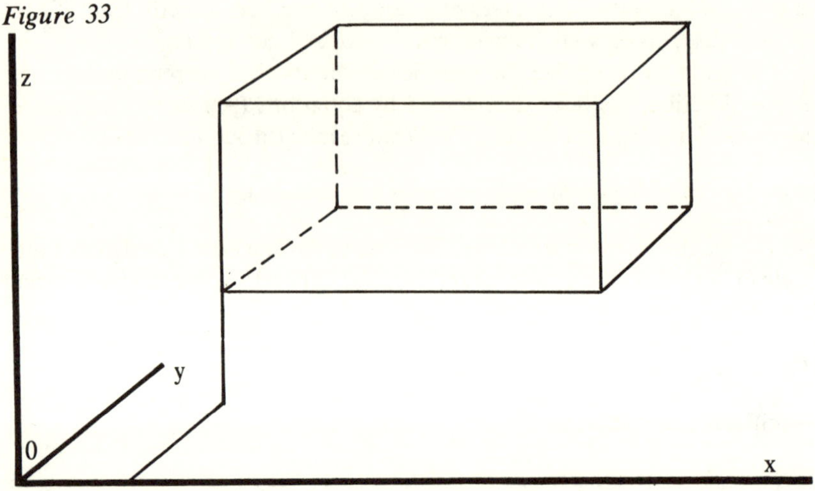

lelepiped during time element dt. And in addition, one has to consider the transit through the six boundary surfaces. There are two side surfaces, each with an area of dy dz, lying perpendicular to the x-axis. In one of these surface areas, all the points have their first coordinate equal to x, and in the other equal to x+dx. The quantity of electricity, $i_1 dy dz dt$, will be flowing through the first side surface during time dt, and the quantity of electricity

$$\left(i_1 + \frac{\partial i_1}{\partial x}\, dx\right) dy\, dz\, dt$$

will be flowing through the second side surface. The first calculated quantity of electricity enters the parallelepiped, and the second calculated quantity leaves it, so that what results for the increase in the interior is

$$-\frac{\partial i_1}{\partial x}\, dx\, dy\, dz\, dt.$$

We will calculate in the same way the increase that originates in the transit through the side surfaces to which the y-axis and the z-axis, respectively, are perpendicular. We obtain

$$-\frac{\partial i_2}{\partial y}\, dx\, dy\, dz\, dt, \quad -\frac{\partial i_3}{\partial z}\, dx\, dy\, dz\, dt.$$

If one adds the last three expressions, what results is the total increase in free electricity that is allotted to the infinitely small parallelepiped during time element dt.

But on the other hand, we have to note that $\varrho\, dx\, dy\, dz$ is all of the free electricity that is found in the parallelepiped at time t. During the next time element dt, this undergoes the increase

$$\frac{\partial \rho}{\partial t}\, dx\, dy\, dz\, dt.$$

Therefore, we have two different expressions for this same increase, and by equating them we get the important equation

$$\frac{\partial \rho}{\partial t} = -\left(\frac{\partial i_1}{\partial x} + \frac{\partial i_2}{\partial y} + \frac{\partial i_3}{\partial z}\right).$$

Section 56

Force Of Separation, Specific Current Intensity, And Specific Resistance

As we saw in section 45, electric equilibrium for an arbitrary conductor consists of having the free electricity distributed over the surface with a definite density for every point, and of having a value of zero for the density of electricity at every place in the interior. However, this condition can also be interpreted (section 44) as consisting of equal quantities of positive and negative electricity present in a *neutral mixture* in every spatial element, no matter how small, in the interior.

If we now consider a single electric particle after *equilibrium has been established,* then there are two interpretations offered for its behavior. One can either assume that every particle is at rest, or one can imagine all the electric particles as being in *permanent* motion, in a motion in which the equilibrium of the electricities consists in not having the *density* undergo a change anywhere in the conductor, while the *specific current intensity* is equal to zero everywhere.

If the conductor that we are considering changes its location in relation to other electrically charged bodies, then the density remains equal to zero everywhere in the interior. As far as the surface is concerned, every new location for the conductor requires a definite new distribution of free electricity which arises in the very same moment the conductor occupies the new location (section 47). When one suddenly interrupts the relative motion of the conductor under consideration and of all the other charged bodies, then the new equilibrium condition of the electric particles is completed in that very same moment (or, nevertheless, after an immeasurably short time). This behavior still needs some explanation.

If one assumes that every single electric particle is at rest in electric equilibrium, then the only way that a new distribution of free electricity comes about when the conductor is moving is by motion of the electric particles. Then, when all the charged bodies are suddenly fixed, the momentary establishment of a new electric equilibrium can only be explained by having the velocity of every electric particle become zero in an immeasurably short time.

The situation is different if one assumes the electric particles as being in motion *under all circumstances.* In electric equilibrium, this motion is constituted in such a way that the specific current is equal to zero everywhere and the density of the free electricity does not undergo a change anywhere. Then, during the relative motion of the charged body, the motion of the electric particles modifies itself in such a way that the specific current intensity is no longer equal to zero everywhere. The specific current intensities that then occur will cause the density to remain zero everywhere in the interior of the conductor under consideration, and will cause a new distribution of free electricity to be present in the surface with every new location of the charged body, as is required by the electrostatic law. Then, when the charged bodies are suddenly fixed, the velocities of the electric particles will not change to zero. They only will change in such a way that the specific current intensity is immediately zero everywhere.

We prefer this second interpretation. We will assume that the electric particles in the interior of every conductor are conceived of as being in permanent, *extraordinarily rapid* motion which only originates from every electric particle being propelled by its immediate neighboring particles. As long as no *separating forces* are added to these molecular effects then the specific current intensity is zero everywhere. But if a separation of the electricity occurs at any place in the interior of the conductor, then this effect consists of one kind of electricity accelerating in a definite direction, while the other kind of electricity decelerates. As a consequence of this, there will no longer be quantities of electricity passing from the negative side to the positive side, going through a surface element whose normal falls in that direction, which are just as large as the quantities going in the reverse direction, from the positive side to the negative side. In other words, *whenever separation occurs, the specific current intensity in the direction of the separating force is no longer equal to zero. But, on the other hand, the specific current remains zero for every direction at right angles to the separating force.*

In comparison to the molecular forces which bring about the electric particles' incessant and very rapid motion, we will have to assume the separating forces to be infinitesimally small. The specific current intensity at any place is therefore a continuous function of that very same separation force that is occurring, a continuous function that becomes zero simultaneously with the separation force. One can think of this function as developed for positive, whole, uneven powers of the separation force, and one can limit it to the first power so long as the ratio of the separation force to the electric molecular forces is infinitesimally small. ***That is, the specific current intensity is proportional to the separation force.***

Thus, if we denote $\epsilon\Pi$ as the total electric separation force at point (x, y, z) in the interior of the conductor, and if we denote $\epsilon\Xi$, ϵH, ϵZ as its components in the direction of the coordinate axes, then these equations are valid:

$$(1)\quad \begin{aligned} w\,i_1 &= \Xi,\; w\,i_2 = \mathrm{H},\; w\,i_3 = \mathrm{Z}, \\ w\,i &= w\sqrt{i_1^2 + i_2^2 + i_3^2} = \Pi. \end{aligned}$$

We will call factor w the *specific resistance* of the conductor, and we will call the reciprocal magnitude $1/w = k$ its *efficiency.* Magnitude w (and consequently k too) is constant for one and the same homogeneous conductor. But the conductor's value is another value, depending on the value of the material from which the conductor is constructed.

Section 57

Constant Current. The Three Equations Which Determine The Conditions For Validity Of V

We want to make the special assumption that the *separating force* is independent of time, and thus for all the points in the interior of the conductor under consideration, it is only a function of spatial coordinates x, y, z.

The components of the separating force that act on the quantity of electricity ϵ at point (x, y, z) will be denoted as $\epsilon\Xi$, ϵH, ϵZ and they will be given as a function of x, y, z in the interior of the conductor. Free electricity will accumulate as a result of the separation. We will denote the potential function orginating from this as V. According to section 45:

$$V = -\sum \frac{\varepsilon}{r}.$$

Therefore, the components of the total electromotive force which will be exerted on the quantity of electricity ϵ present at point (x, y, z) are, respectively,

$$\varepsilon\left(\frac{\partial V}{\partial x} + \Xi\right),\; \varepsilon\left(\frac{\partial V}{\partial y} + \mathrm{H}\right),\; \varepsilon\left(\frac{\partial V}{\partial z} + \mathrm{Z}\right),$$

and these components have the directions of the x, y, and z axes respectively. We will obtain the following expressions corresponding to the equations in (1) in section 56:

$$i_1 = k\left(\frac{\partial V}{\partial x} + \Xi\right),\ i_2 = k\left(\frac{\partial V}{\partial y} + \mathrm{H}\right),\ i_3 = k\left(\frac{\partial V}{\partial z} + \mathrm{Z}\right)$$

for the specific current intensity in these same directions. The accumulated free electricity acts against an unlimited new accumulation of free electricity. As a consequence, the density of free electricity any place in the conductor will no longer change after a definite point in time; that is, from this point on in time $\partial\varrho/\partial t=0$. So, by referring to equation (1) in section 55, what we obtain for every point in the *interior* of the conductor is

$$\frac{\partial i_1}{\partial x} + \frac{\partial i_2}{\partial y} + \frac{\partial i_3}{\partial z} = 0$$

or, if the previous expressions for i_1, i_2, i_3 are used,

$$\text{(1)} \quad \frac{\partial\left\{k\left(\frac{\partial V}{\partial x} + \Xi\right)\right\}}{\partial x} + \frac{\partial\left\{k\left(\frac{\partial V}{\partial y} + \mathrm{H}\right)\right\}}{\partial y} + \frac{\partial\left\{k\left(\frac{\partial V}{\partial z} + \mathrm{Z}\right)\right\}}{\partial z} = 0.$$

Let the conductor's free surface be insulated. Then, in every one of the surface's points, the specific current intensity directed perpendicular to the surface is equal to zero. We will then draw a normal from a point (x, y, z) in the free surface towards the interior, and we will denote an interval on it as n. The specific current intensity perpendicular to the surface is

$$-\left(i_1\frac{\partial x}{\partial n} + i_2\frac{\partial y}{\partial n} + i_3\frac{\partial z}{\partial n}\right).$$

If one sets this equation equal to zero, then one obtains the following conditions for every point in the free surface:

$$\text{(2)} \quad \left(\frac{\partial V}{\partial x} + \Xi\right)\frac{\partial x}{\partial n} + \left(\frac{\partial V}{\partial y} + \mathrm{H}\right)\frac{\partial y}{\partial n} + \left(\frac{\partial V}{\partial z} + \mathrm{Z}\right)\frac{\partial z}{\partial n} = 0.$$

If magnitudes Ξ, H, Z, k do not change discontinuously any place inside the conductor, then no more equations will occur. However, we still want to consider the case where those magnitudes change discontinuously by transition through single surfaces in the conductor's interior. Let a normal be drawn toward both sides from a point (x, y, z) of the surface of discontinuity. Then, an interval p calculated on it as positive on one side and negative on the other. What is still to be expressed is that two points on this normal, which lie infinitely close to point (x, y, z) on the different sides of the surface, will have the same specific current intensity in the direction of increasing p. Or, in other words, just as much electricity will flow into the space on the positive side in each time element at any place on the surface of discontinuity as flows out of the space on the negative side. This is expressed in the equation

$$\text{(3)} \quad \begin{aligned} &\left\{k\left(\frac{\partial V}{\partial x}+\Xi\right)\frac{\partial x}{\partial p}+k\left(\frac{\partial V}{\partial y}+\mathrm{H}\right)\frac{\partial y}{\partial p}+k\left(\frac{\partial V}{\partial z}+\mathrm{Z}\right)\frac{\partial z}{\partial p}\right\}_{p=-0} \\ =&\left\{k\left(\frac{\partial V}{\partial x}+\Xi\right)\frac{\partial x}{\partial p}+k\left(\frac{\partial V}{\partial y}+\mathrm{H}\right)\frac{\partial y}{\partial p}+k\left(\frac{\partial V}{\partial z}+\mathrm{Z}\right)\frac{\partial z}{\partial p}\right\}_{p=+0}. \end{aligned}$$

Equation (3) is valid for every point (x, y, z) of the surface of discontinuity in the interior of the conductor.

Section 58

The Unique Existence Of V

It will now be proven that there always exists a function V which fulfills conditions (1), (2), (3) in section 57 and which has values of a given difference on both sides of every surface of discontinuity. In order to perform this proof, we will exclude from the space which the conductor fills those infinitely thin spaces that contain the surfaces of discontinuity. Section 21 discusses how this is done, and Figure 12 explains it. The remaining space in the conductor is denoted by S. In its interior, Ξ, H, Z, k will be finite and free of discontinuities everywhere. By v is understood any function of x, y, z which fulfills both of the following conditions. The values for v should have a given finite difference for any two points which lie infinitely close to each other on the opposite sides of a surface of discontinuity, and the first derivative of v should be finite and continuously variable everywhere in the interior of space S.

There are infinitely many of these v functions. If one of them is designated as V, then all the others can be expressed in the form v=V+hs, if h is a constant that is appropriate for selection and if s is a function for x, y, z that satisfies the same conditions as v does, but which itself does *not* become discontinuous in the surfaces of discontinuity of v.

According to this, the integral

$$\text{(1)} \quad \Omega(v)=\int k\left\{\left(\frac{\partial v}{\partial x}+\Xi\right)^2+\left(\frac{\partial v}{\partial y}+\mathrm{H}\right)^2+\left(\frac{\partial v}{\partial z}+\mathrm{Z}\right)^2\right\}dS,$$

extended over space S, has a *finite, positive* value. This value changes when one changes over from one function v to another. Therefore, among all of the permissible functions v, there is a least one — we will want to designate it as V — which turns the value of the integral into a *minimum*. The condition for this is as follows:

$$(2)\quad \Omega(V) < \Omega(V + hs),$$

if h is taken to be infinitely small. And, we can now develop $\Omega(V+hs)$. The calculation procedure is prescribed in section 34. One obtains

$$\begin{aligned}
&\Omega(V+hs)\\
&= \Omega(V)\\
(3)\quad &+ 2h\int k\left\{\left(\frac{\partial V}{\partial x}+\Xi\right)\frac{\partial s}{\partial x}+\left(\frac{\partial V}{\partial y}+\mathrm{H}\right)\frac{\partial s}{\partial y}+\left(\frac{\partial V}{\partial z}+\mathrm{Z}\right)\frac{\partial s}{\partial z}\right\}dS\\
&+ h^2\int k\left\{\left(\frac{\partial s}{\partial x}\right)^2+\left(\frac{\partial s}{\partial y}\right)^2+\left(\frac{\partial s}{\partial z}\right)^2\right\}dS.
\end{aligned}$$

The first and third components on the right-hand side of equation (3) are positive. The second can turn out to be positive as well as negative. In order for condition (2) to be satisfied, it is necessary and sufficient that

$$(4)\quad \int k\left\{\left(\frac{\partial V}{\partial x}+\Xi\right)\frac{\partial s}{\partial x}+\left(\frac{\partial V}{\partial y}+\mathrm{H}\right)\frac{\partial s}{\partial y}+\left(\frac{\partial V}{\partial z}+\mathrm{Z}\right)\frac{\partial s}{\partial z}\right\}dS = 0.$$

For, in actuality, a *positive* term can be added to Ω (V) on the right-hand side of (3) which will then become zero only when s=const. everywhere. So, equation (4) is sufficient for the occurrence of equation (2). But it is also necessary. For if it were not satisfied, then one could select the sign for h in such a way that the second component on the right-hand side of (3) would turn out to be negative, and one could make the numerical value for h so small that the third component would become smaller than the numerical value of the second component. And then one would have

$$\Omega(V+hs) < \Omega(V),$$

which is in contradiction to (2).

We can now transform the integral on the left-hand side of equation (4) according to section 20. As a result, equation (4) transforms into the following:

$$\begin{aligned}
(5)\quad &-\int s\left\{\frac{\partial\left\{k\left(\frac{\partial V}{\partial x}+\Xi\right)\right\}}{\partial x}+\frac{\partial\left\{k\left(\frac{\partial V}{\partial y}+\mathrm{H}\right)\right\}}{\partial y}+\frac{\partial\left\{k\left(\frac{\partial V}{\partial z}+\mathrm{Z}\right)\right\}}{\partial z}\right\}dS\\
&-\int k\,s\left\{\left(\frac{\partial V}{\partial x}+\Xi\right)\frac{\partial x}{\partial n}+\left(\frac{\partial V}{\partial y}+\mathrm{H}\right)\frac{\partial y}{\partial n}+\left(\frac{\partial V}{\partial z}+\mathrm{Z}\right)\frac{\partial z}{\partial n}\right\}d\sigma\\
&= 0.
\end{aligned}$$

The first of these two integrals is to be extended over the entire space S, and the second is to be extended over its total surface. If equation (5) is to be satisfied, then one has to have both integrals equal *zero simultaneously*. The spatial integral becomes zero when for every point (x, y, z) in the interior of S, the magnitude in parentheses multiplied by s dS has the value of zero. *This gives equation (1) of section 57.*

The surface of S first of all consists of the free surface of the conductor, and second of the boundaries around the surfaces of discontinuity in the interior. So one first has to have what is multiplied by ks dσ equal to zero for every point (x, y, z) in the free surface of the conductor. *This gives equation (2) of section 57.*

The boundaries of a surface of discontinuity are two surfaces which lie infinitely close to it on two opposing sides and which produce infinitely small segments of p=-ε and p=+ε on its respective normals. Because normal n is always drawn toward the interior of space S, n=p on the positive side of the surface of discontinuity, and n=-p on the negative side. Function s changes continuously when point (x, y, z) goes through the surface of discontinuity. Thus, s has two values for two points which, lying on the negative and the positive side of this normal lie infinitely close to the surface. These two values only differ infinitely slightly from the value at the starting point of the normal. According to this, then, the surface integral which is to be extended over both of the boundaries of a surface of discontinuity is written in this way:

$$-\int d\sigma . s . \left\{ \begin{array}{l} \left[k\left(\frac{\partial V}{\partial x}+\Xi\right)\frac{\partial x}{\partial p}+k\left(\frac{\partial V}{\partial y}+\mathrm{H}\right)\frac{\partial y}{\partial p}+k\left(\frac{\partial V}{\partial z}+\mathrm{Z}\right)\right]_{p=+0} \\ -\left[k\left(\frac{\partial V}{\partial x}+\Xi\right)\frac{\partial x}{\partial p}+k\left(\frac{\partial V}{\partial y}+\mathrm{H}\right)\frac{\partial y}{\partial p}+k\left(\frac{\partial V}{\partial z}+\mathrm{Z}\right)\right]_{p=-0} \end{array} \right\}.$$

This integral is to be extended over all the surfaces of discontinuity once as a contribution to equation (5). In order for this integral to have a value of zero, one has to have what is multiplied by s dσ in the last integral equal to zero for every point (x, y, z) in all surfaces of discontinuity. *This gives the equation (3) in section 57.*

Since at least one of the functions among all the permissible functions v turns integral (1) into a minimum, this *one* function V satisfies conditions (1), (2), and (3) of section 57. In addition to that, it can be shown that, disregarding an additive constant, this function V is the sole solution for the problem. Assuming that in addition to V there is still another function V+s which also turns integral (1) into a minimum, then the conditions for this would be

$$(6) \qquad \Omega(V+s) \leqq \Omega(V+hs),$$

if we now understand h to be a constant that lies infinitely close to one.

But, if we also consider that function V satisfies equation (4), then what results is

$$\Omega(V+hs) = \Omega(V) + h^2 \int k\left\{\left(\frac{\partial s}{\partial x}\right)^2 + \left(\frac{\partial s}{\partial y}\right)^2 + \left(\frac{\partial s}{\partial z}\right)^2\right\} dS$$

and

$$\Omega(V+s) = \Omega(V) + \int k\left\{\left(\frac{\partial s}{\partial x}\right)^2 + \left(\frac{\partial s}{\partial y}\right)^2 + \left(\frac{\partial s}{\partial z}\right)^2\right\} dS.$$

Consequently, condition (6) turns into the following form:

$$(7)\quad \int k\left\{\left(\frac{\partial s}{\partial x}\right)^2 + \left(\frac{\partial s}{\partial y}\right)^2 + \left(\frac{\partial s}{\partial z}\right)^2\right\} dS \leqq h^2 \int k\left\{\left(\frac{\partial s}{\partial x}\right)^2 + \left(\frac{\partial s}{\partial y}\right)^2 + \left(\frac{\partial s}{\partial z}\right)^2\right\} dS.$$

One may, however, assume that the constant h^2, which lies infinitely close to one, is simply larger than one, or that it is smaller than one too. Therefore, condition (7) is only satisfied provided that the integral has a value of zero. In order to obtain this, one has to have

$$\frac{\partial s}{\partial x} = 0, \quad \frac{\partial s}{\partial y} = 0, \quad \frac{\partial s}{\partial z} = 0$$

for every point in the conductor's interior, i.e., s=const.

So there is always a function V which satisfies conditions (1), (2), and (3) and which has values of a given difference for both sides of every interior surface of discontinuity. Every other function that also does this is distinguished from the former only by an additive constant.

The value of the additive constant cannot be determined from the given separating forces. If a point on the conductor's surface is connected to the earth by an infinitely thin wire, then V=0 at this point. On the other hand, if the conductor is totally insulated, then the algebraic sum of the quantities of electricity is constant, and naturally equal to zero, if no free electricity was originally present. In both cases, this gives a secondary condition for determining the additive constant.

After function V has been determined for the interior and the surface of the conductor, then what is necessary is to extend it into the *exterior* space *continuously* so that it will satisfy Laplace's equation

$$\frac{\partial^2 V}{\partial x^2} + \frac{\partial^2 V}{\partial y^2} + \frac{\partial^2 V}{\partial z^2} = 0$$

for every point there, and so that it will be equal to zero at an infinite distance. It was already shown in section 34 that this continuation of function V always occurs only in one way. In order to determine it, Green's Theorem must be used (section 21).

If function V is known for every point in all of infinite space, then the density of the free electricity in the interior of the conductor can be found according to formula (6), and, on the surface, can be found according to formula (7), in section 45. If k=const. and $\Xi = \mathrm{H} = \mathrm{Z} = 0$ in a section of the conductor, then equation (1) in section 57 transforms into

$$\frac{\partial^2 V}{\partial x^2} + \frac{\partial^2 V}{\partial y^2} + \frac{\partial^2 V}{\partial z^2} = 0,$$

i.e., the density of electricity in this section of the conductor is equal to zero.

Section 59

The Work Performed By Electricity In Motion

We still want to investigate the significance of integral (1) of section 58 in the case where its value is a minimum. This integral is

$$\text{(1)} \quad \int k\left\{\left(\frac{\partial V}{\partial x} + \Xi\right)^2 + \left(\frac{\partial V}{\partial y} + \mathrm{H}\right)^2 + \left(\frac{\partial V}{\partial z} + \mathrm{Z}\right)^2\right\} dS.$$

Function V, which yields the minimal value, has been proven to be the potential function of free electricity. Therefore, the following equations are valid (section 57):

$$\text{(2)} \quad w\, i_1 = \frac{\partial V}{\partial x} + \Xi, \; w\, i_2 = \frac{\partial V}{\partial y} + \mathrm{H}, \; w\, i_3 = \frac{\partial V}{\partial z} + \mathrm{Z}.$$

And with their help expression (1) can be written as

$$\text{(3)} \quad \int\left\{i_1\left(\frac{\partial V}{\partial x} + \Xi\right) + i_2\left(\frac{\partial V}{\partial y} + \mathrm{H}\right) + i_3\left(\frac{\partial V}{\partial z} + \mathrm{Z}\right)\right\} dS$$

or even

$$\text{(4)} \quad \int w\,(i_1^2 + i_2^2 + i_3^2)\, dS$$

or finally

$$\text{(5)} \quad \int dS\,.\, w\, i^2.$$

It is easy to recognize the mechanical significance of these expressions. Because of equations (5) in section 54, one can also write

$$\text{(6)} \quad \sum \varepsilon\left\{\left(\frac{\partial V}{\partial x} + \Xi\right)\frac{dx}{dt} + \left(\frac{\partial V}{\partial y} + \mathrm{H}\right)\frac{dy}{dt} + \left(\frac{\partial V}{\partial z} + \mathrm{Z}\right)\frac{dz}{dt}\right\}$$

instead of (3). This summation is based on all the electric particles of the entire conductor. If one designates A as the work performed by the electricity in motion at time t, then expression (6) is nothing else than dA/dt. Thus we have

$$(7)\quad \int dS\,.\,w\,i^2 = \frac{dA}{dt}.$$

Section 60

Special Case: Separation Only Occurs In An Infinitely Thin Layer

We will now consider the special case where the electromotive forces $\epsilon\Xi$, ϵH, ϵZ only occur in an infinitely thin layer, for example, on the tangential surface between two heterogeneous, component parts of the conductor. In this case, the condition that V must satisfy in the interior of the conductor system can be transformed even further. This condition reads so that integral

$$(1)\quad \int k\left\{\left(\frac{\partial V}{\partial x}+\Xi\right)^2+\left(\frac{\partial V}{\partial y}+\mathrm{H}\right)^2+\left(\frac{\partial V}{\partial z}+\mathrm{Z}\right)^2\right\} dS$$

expanded over an entire conductor system must be a minimum, and when this condition is satisfied, then integral (1) has the value

$$(2)\quad \int dS\,.\,w\,i^2.$$

Naturally there are infinitely many V functions that satisfy this condition. But each pair of them will have a constant difference everywhere in the interior of the conductor. Consequently they will all yield *one and the same* minimal value (2) of integral (1). And one will obtain a deviation from this minimal value when one replaces

$$\frac{\partial V}{\partial x},\ \frac{\partial V}{\partial y},\ \frac{\partial V}{\partial z}$$

by zero everywhere. However, because only *one* minimal value of integral (1) exists, every deviation must turn out to be larger than the true minimal value (2). That is, we get the inequality

$$(3)\quad \int k(\Xi^2+\mathrm{H}^2+\mathrm{Z}^2)\,dS > \int dS\,.\,w\,i^2.$$

But according to our assumption, components Ξ, H, Z are only different from zero in an infinitely thin layer. What will then drop out of the left-hand side of (3) are all those contributions which belong to spatial elements that are outside of that layer. For the integral

$$\int k(\Xi^2 + H^2 + Z^2)\, dS$$

there only remains one domain of integration which is infinitely small in comparison to the space over which integral (2) is to be extended. It follows from this that inequality (3) cannot be satisfied in any other way than when the value that $\Xi^2 + H^2 + Z^2$ has in the boundary layer is infinitely large in comparison to i^2. Therefore, equation (2) of section 59 will change into

$$(4) \quad \frac{\partial V}{\partial x} = -\Xi, \quad \frac{\partial V}{\partial y} = -H, \quad \frac{\partial V}{\partial z} = -Z$$

for this layer.

Then, we will construct the normal to both sides at an arbitrary point on the surface which separates the two heterogeneous conductor components, counting off intervals p positive on one side, negative on the other. We will then choose one point each on the negative and positive normals, and infinitely close to the separation surface. Let the first point have coordinates x, y, z; then the other has coordinates

$$x + \frac{\partial x}{\partial p} dp, \quad y + \frac{\partial y}{\partial p} dp, \quad z + \frac{\partial z}{\partial p} dp,$$

and dp is the distance between them. If we then multiply equations (4) on both sides by

$$\frac{\partial x}{\partial p} dp, \quad \frac{\partial y}{\partial p} dp, \quad \frac{\partial z}{\partial p} dp,$$

respectively, and add this, then what results is

$$(5) \quad V_{+0} - V_{-0} = -\left(\Xi \frac{\partial x}{\partial p} + H \frac{\partial y}{\partial p} + Z \frac{\partial z}{\partial p}\right) dp.$$

Here, V_{-0} and V_{+0} denote the values of function V in each of the two points lying infinitely close to the separation surface. The difference of these values is finite, and for each point on the separation surface it is known since Ξ, H, Z are given everywhere in the infinitely thin boundary layer. And so the assumption in section 58 applies here, that the difference in the values for V is given for any two points which lie infinitely close to each other on opposite sides of the surface of discontinuity. Outside the boundary layer, function V changes continuously in the remaining space filled by the conductor system. $\Xi = \mathrm{H} = \mathrm{Z} = 0$ in this remaining space. As a result, the condition for V is now simpler:

$$(6) \quad \int k \left\{ \left(\frac{\partial V}{\partial x}\right)^2 + \left(\frac{\partial V}{\partial y}\right)^2 + \left(\frac{\partial V}{\partial z}\right)^2 \right\} dS = Min.,$$

if $V_{+0} - V_{-0}$ is given for the separation surface.

If one has now determined V for the interior of the conductor, then the specific current intensities in the direction of the coordinate axes will result from these equations:

$$(7) \quad w\, i_1 = \frac{\partial V}{\partial x},\; w\, i_2 = \frac{\partial V}{\partial y},\; w\, i_3 = \frac{\partial V}{\partial z}.$$

It was Ohm† who first dealt with this case. He called function V the *voltage.* But he was mistaken about this function's significance, for he believed that it expressed the density of the electricity.

The difference in voltages $V_{+0} - V_{-0}$ on both sides of the boundary surface between two heterogeneous conductor components depends on the nature of both of these components. If the difference in voltage is known for the boundary surface (or, when there are several boundary surfaces present, if all have the same difference in voltage), then function V is uniquely determined up to an additive constant by condition (6). It has already been discussed in section 58 how to determine this constant's value and how function V can be extended continuously into exterior space in only one way.

Section 61

Further Specific Cases: Wire-Shaped Conductors. Ohm's Law

Let a section of the conductor system have the shape of a wire. What we mean by a *wire* is a body in whose interior we can draw a continuous line (the axis) so that every planar cross-section q placed perpendicular to it will have infinitesimally small dimensions in comparison to the length of the axis. On the axis, s denotes an arc running from the beginning point to an indefinite point. Naturally, cross-section q does not need to be the same everywhere. Nevertheless, we will stipulate that given a continuous change in s, the changes in the cross-section will proceed continuously in such a way that one can make two cross-sections sufficiently close to each other at any place. They will deviate from each other, and from all cross-

† *Ohm, G.S.* Die galvanische Kette, mathematisch behandelt. Berlin 1827.

sections lying in between, only by an infinitely small amount. Between each such two cross-sections, the wire can be regarded as a cylinder of arbitrarily shaped, but unchanging cross-section.

First of all, we will only consider *one* such cylinder at an arbitrary place on the wire. This cylinder's axis will be in finite ratio to the dimensions of cross-section q. We can therefore view the axis as rectilinear and place the x-axis of the rectangular coordinate system on it. Thus, the perpendicular cross-section will be parallel to the yz plane. Because the dimensions of every cross-section are so infinitely small, we can neglect the current in its plane in comparison to the current that is directed perpendicularly against this plane. That is, *in every direction in which the plane of a cross-section falls, we are allowed to have the specific current intensity equal to zero.*

(1) $i_2 = 0, \quad i_3 = 0.$

Furthermore, in *one and the same* cross-section, we are allowed to take each of the components Ξ, H, Z, and the specific resistance w as constant. However, it now follows from (1) and from equations (2) in section 59 that

$$\frac{\partial V}{\partial y} = -\mathrm{H}, \quad \frac{\partial V}{\partial z} = -\mathrm{Z},$$

i.e., for every point inside the same cross-section $V = -\mathrm{H}y - \mathrm{Z}z + f(x)$. But, because y and z are infinitely small in every cross-section, one has a shorter expression $V = f(x)$. According to this, the first of the equations in (2) in section 59 gives

(2) $w\, i_1 = f'(x) + \Xi.$

Now $f'(x)$ is only dependent on x, just as Ξ, according to our assumption, has the same value in all the points of the same cross-section. *Consequently, for one and the same cross-section, the specific current intensity that is perpendicularly directed against it is constant in all of its points.*

After this preparation we will now consider the wire-shaped conductor in its entire extension. We will make a cross-section q perpendicular to the axis, which will cut off an arc length s on the same axis; dq is a surface element of the cross-section, which should be considered as the base of a spatial element, whose height is ds. Therefore, the volume of this spatial element is $dS = dq\, ds$.

Then, if we once again denote A as the work performed by the electricity in motion at time t, according to section 59, equation (7), we get:

(3) $$\frac{dA}{dt} = \int\int dq\, ds\, w\, i^2 = \int\int dq\, ds\, i\left(\frac{\partial V}{\partial s} + \Pi\right).$$

V here is the potential function of free electricity and Π is the component of the electromotive force taken in the s direction.

According to the explanation of *specific current intensity,* i dq dt is the algebraic sum of the quantities of electricity that go from the negative to the positive side of cross-section dq at place s during time element dt reduced by the algebraic sum of those quantities which go through in the opposite direction at the same place during the same time element. For all of cross-section q, the difference in question amounts to i q dt, which we will want to denote as J dt. Thus

$$(4) \quad J = i\, q.$$

We will call J the *current intensity at cross-section q.* The axis, on which arc s is counted, lies perpendicular to all cross-section. Therefore, in every cross-section we can allow the *positive* normal to coincide *with the direction of increasing arc s.* Because we will now assume that there is no more accumulation of free electricity occurring, or rather that a state of resistance has commenced, J must have the same value for all the places in the wire, or — what is the same— J is independent of s. Equation (3) changes into the following:

$$\frac{dA}{dt} = \int w \frac{J^2}{q}\, ds = \int J \left(\frac{\partial V}{\partial s} + \Pi \right) ds$$

and according to the last comment made about J this equation can also be written as

$$(5) \quad \frac{dA}{dt} = J^2 \int w \frac{ds}{q} = J \int \left(\frac{\partial V}{\partial s} + \Pi \right) ds.$$

We will have

$$\int w \frac{ds}{q} = W, \quad \int \Pi\, ds = E$$

and will call W the *resistance* of a wire-shaped conductor and E the *integral value of the electromotive force.* If we designate V_1 and V_2, respectively, as the values that function V has at the beginning and end points of the wire-shaped conductor, then what results from equation (5) is

$$(6) \quad \frac{dA}{dt} = J^2 \, . \, W = \{ E + (V_2 - V_1) \} \, . \, J$$

in a looping wire-shaped conductor $V_2 = V_1$, and, consequently, the result here is

$$(7) \quad \frac{dA}{dt} = J^2 \, . \, W = E \, . \, J$$

and, therefore, in a closed linear current

$$(8)\quad J = \frac{E}{W}.$$

The relationship that is expressed in this equation between the current intensity, the resistance, and the integral value of the electromotive force will be called Ohm's Law.

Section 62

Continuation: Branched Wires

If the conductor system is composed of arbitrarily wire-shaped branches, then we will apply equation (6) of section 61 to every segment between two nodal points. Let the nodal points be numbered. The value that function V will have in any one of these nodal points will be designated by having the number of the point attached to V as an index. We will attach two indices to J, E, and W, which will specify the beginning and end points, respectively, of the branch to which the values for J, E, and W belong. Then generally,

$$J_{h,k} = -J_{k,h}, \quad E_{h,k} = -E_{k,h}, \quad W_{h,k} = W_{k,h}.$$

Figure 34

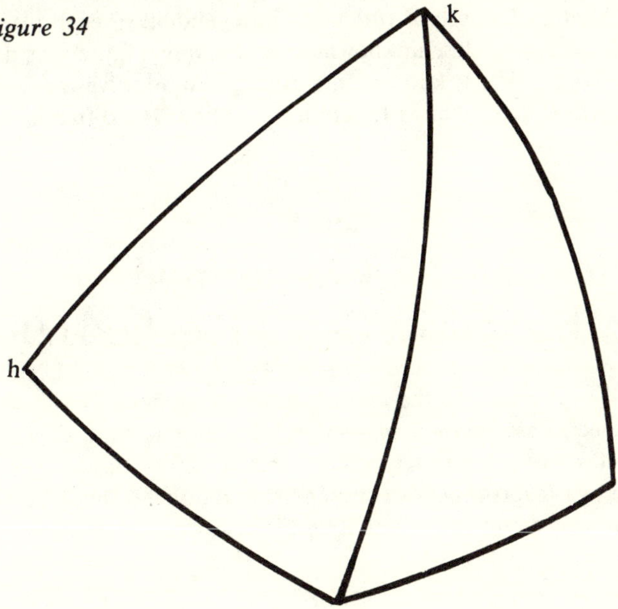

The number of the unbranched components of the conductor system will be m, and n will be the number of nodal points. Therefore, we have m number of equations of the form

(1) $J_{h,k}\, W_{h,k} = V_k - V_h + E_{h,k}.$

We still, moreover, have to express the fact that there is no accumulation of electricity occurring in any nodal point. Thus, at any time, the quantity of electricity entering the nodal point in the next time element dt must be just as large as the quantity of electricity leaving the nodal point during that same time element. Or, in other words, the *algebraic sum* of the current intensities in all of the branches which emanate from the nodal point, going away from it in every direction, must be equal to zero. For example, if the only branches that emanate from nodal point 1 are 1,2; 1,3; 1,4, and no others, then we get

(2) $J_{1,2} + J_{1,3} + J_{1,4} = 0.$

There are n number of equations existing with this form. But one of them is an identical result of the n-1 remaining ones. And, if $J_{h.k} = -J_{k.h}$, then one will get an identical 0=0 if one combines all of the equations in (2) by addition.

The values for E and W are known for every branch in the conductor system, but, on the other hand, V and J are unknown. So m+n will be the number of these unknowns. Since m+n-1 is the number of linear equations (1) and (2), both of which are independent of each other, one can uniquely calculate these unknowns from the linear equations if one of them, for example, V_1, is known. In actuality, an additive constant with an arbitrary value can occur in function V, without the differences in (1) changing.

Section 63

The Work Performed In The Special Case In Section 60

We will now return to the investigations in sections 57, 58, and 59, using the special assumption in section 60, namely, that electromotive forces $\epsilon\Xi$, ϵH, ϵZ will only occur in the infinitely thin boundary layer at the point where two heterogeneous component parts of the conductor touch. According to this assumption, equation (7) in section 59 changes into the following:

(1) $$\frac{dA}{dt} = \int\int\int \left\{ i_1 \frac{\partial V}{\partial x} + i_2 \frac{\partial V}{\partial y} + i_3 \frac{\partial V}{\partial z} \right\} dx\, dy\, dz.$$

In this equation, dxdydz is a spatial element in the interior of the conductor and the integration is to be extended over all of the space filled by the conductor. But now we also have the identical equations

$$i_1 \frac{\partial V}{\partial x} = \frac{\partial (i_1 V)}{\partial x} - V \frac{\partial i_1}{\partial x}, \quad i_2 \frac{\partial V}{\partial y} = \frac{\partial (i_2 V)}{\partial y} - V \frac{\partial i_2}{\partial y},$$

$$i_3 \frac{\partial V}{\partial z} = \frac{\partial (i_3 V)}{\partial z} - V \frac{\partial i_3}{\partial z}.$$

Therefore, equation (1) can be transformed. We obtain

$$\begin{aligned} \frac{dA}{dt} &= -\iiint V \left(\frac{\partial i_1}{\partial x} + \frac{\partial i_2}{\partial y} + \frac{\partial i_3}{\partial z} \right) dx\, dy\, dz \\ &\quad - \int d\sigma \,.\, V \left(i_1 \frac{\partial x}{\partial n} + i_2 \frac{\partial y}{\partial n} + i_3 \frac{\partial z}{\partial n} \right). \end{aligned} \tag{2}$$

In this equation, the first integral extends over all the space filled by the conductor, and the second integral extends over its total surface, that is, over the insulated, free surface and over the closures around the surfaces of discontinuities. Here, these surfaces of discontinuity are the surfaces in which any two heterogeneous parts of the conductor components touch; n denotes the normal drawn from surface element $d\sigma$ towards the conductor's interior.

For the stationary state which we are assuming, we have for each point in the interior of the conductor,

$$\frac{\partial i_1}{\partial x} + \frac{\partial i_2}{\partial y} + \frac{\partial i_3}{\partial z} = 0 \tag{3}$$

every place in the insulated free surface.

$$i_1 \frac{\partial x}{\partial n} + i_2 \frac{\partial y}{\partial n} + i_3 \frac{\partial z}{\partial n} = 0. \tag{4}$$

Consequently, only the surface integral still remains on the right-hand side of equation (2), which extends over both sides of each surface of discontinuity:

$$\frac{dA}{dt} = -\int d\sigma \; V \left(i_1 \frac{\partial x}{\partial n} + i_2 \frac{\partial y}{\partial n} + i_3 \frac{\partial z}{dn} \right). \tag{5}$$

We will next take a surface element $d\sigma$ in any surface of discontinuity and construct a normal going toward both sides as one of the element's points and count off p intervals from its base outwards, positive on one side and negative on the other. Then, n=p on the side of the positive normal, and n=-p on the side of the negative normal. Consequently, the following can be written instead of equation (5):

$$(6)\quad \frac{dA}{dt} = -\int d\sigma \cdot \left\{ \begin{array}{l} \left[V\left(i_1 \frac{\partial x}{\partial p} + i_2 \frac{\partial y}{\partial p} + i_3 \frac{\partial z}{\partial p}\right)\right]_{p=+0} \\ -\left[V\left(i_1 \frac{\partial x}{\partial p} + i_2 \frac{\partial y}{\partial p} + i_3 \frac{\partial z}{\partial p}\right)\right]_{p=-0} \end{array} \right\}.$$

The integration is to be extended once over every surface of discontinuity in this equation. But, if we take equation (3) in section 57 into consideration, then what immediately results is

$$\left(i_1 \frac{\partial x}{\partial p} + i_2 \frac{\partial y}{\partial p} + i_3 \frac{\partial z}{\partial p}\right)_{p=-0}$$
$$= \left(i_1 \frac{dx}{\partial p} + i_2 \frac{\partial y}{\partial p} + i_3 \frac{dz}{\partial p}\right)_{p=+0}$$

and therefore, instead of (6), one obtains a simpler equation

$$(7)\quad \frac{dA}{dt} = -\int d\sigma\,(V_{+0} - V_{-0})\left(i_1 \frac{\partial x}{\partial p} + i_2 \frac{\partial y}{\partial p} + i_3 \frac{\partial z}{\partial p}\right).$$

Here too, the integration is to be extended over every surface of discontinuity only once. If we now set up the special assumption that the product of the *electromotive forces* $\epsilon\Xi$, ϵH, ϵZ *is constant and is directed perpendicular to the surface* in all points of one and the same surface of discontinuity, then the voltage difference for every one of these surfaces is $V_{+0}-V_{-0}$=const., furthermore, for an arbitrary surface of discontinuity, it is

$$\int d\sigma \left(i_1 \frac{\partial x}{\partial p} + i_2 \frac{\partial y}{\partial p} + i_3 \frac{\partial z}{\partial p}\right) = J,$$

where J signifies the *current intensity* in the direction of the increasing p. Jdt, then, is the quantity of electricity which goes through the surface of discontinuity in the indicated direction in time element dt. According to this, equation (7) can be simplified into the following:

$$(8)\quad \frac{dA}{dt} = -\sum J\,(V_{+0} - V_{-0}),$$

in which sign Σ is to signify that the product $J(V_{+0}-V_{-0})$ is formed for every surface of discontinuity, and that all the products should then be summed.

We will then place two sectional planes σ_1 and σ_2 across the conductor, which will cut out from it a totally bounded piece. Surface σ_1 should simply connected. Its boundary should consist of a single loop which does not cut through itself, and which simultaneously lies in the conductor's

Figure 35

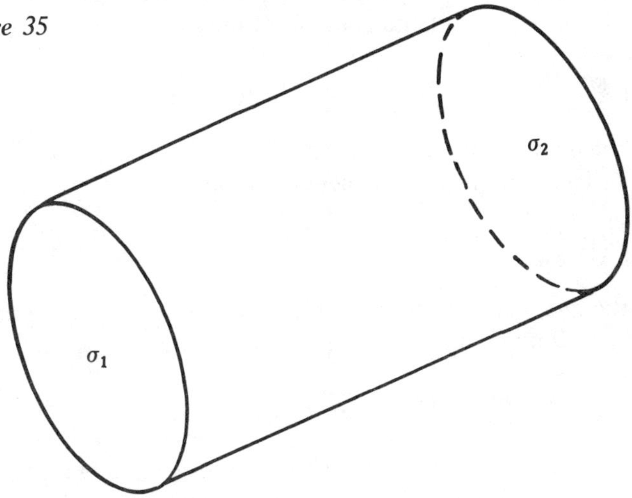

free surface. The same goes for σ_2. [Figure 35] Furthermore, σ_1 and σ_2 will be equipotential surfaces of the potential function, i.e., the potential function should have the same constant value V_1 in all the points in σ_1 and the same constant value V_2 in all the points in σ_2. Then the question here is to calculate dA/dt for the piece of the conductor lying between σ_1 and σ_2. Here, the integral (1) is to be extended over just this piece of the conductor, and, similarly, the volume integral in (2) is to be extended. On the other hand, the surface integral in (2) is to be extended over the insulated free surface between the boundary lines for σ_1 and σ_2, over σ_1 and σ_2, and over the envelopes around the discontinuous surfaces that possibly are present in the piece of conductor. Once again, the volume integral in (2) is equal to zero, as is the surface integral extended over the insulated free surface. The transformation contained in equations (5), (6), (7), and (8) are valid for the envelopes around the surfaces of discontinuity. There is also a contribution to integral (1) coming from the surfaces of discontinuity, $^{-}\Sigma J(V_{+0}-V_{-0})$ in which the summation is only based on the surfaces present in the piece of conductor. The surface integral extended over σ_1 and σ_2 are the only items remaining. Because V_1 and V_2 are constant, these integrals can be written

$$-V_1\int\left(i_1\frac{\partial x}{\partial p}+i_2\frac{\partial y}{\partial p}+i_3\frac{\partial z}{\partial p}\right)d\sigma_1$$

$$+V_2\int\left(i_1\frac{\partial x}{\partial p}+i_2\frac{\partial y}{\partial p}+i_3\frac{\partial z}{\partial p}\right)d\sigma_2.$$

The direction of positive p is placed here in such a way that it goes into the piece of conductor from σ_1 and goes out of the conductor from σ_2. So now

$$dt \int \left(i_1 \frac{\partial x}{\partial p} + i_2 \frac{\partial y}{\partial p} + i_3 \frac{\partial z}{\partial p} \right) d\sigma_1 = J_1 \, dt,$$

which is the quantity of electricity which flows into the conductor system through surface σ_1 in the time element dt, as opposed to

$$dt \int \left(i_1 \frac{\partial x}{\partial p} + i_2 \frac{\partial y}{\partial p} + i_3 \frac{\partial z}{\partial p} \right) d\sigma_2 = J_2 \, dt,$$

which is the quantity of electricity which goes out through σ_2 in that same time element. Both quantities are equal to each other. So, we get

$$(9) \quad \frac{dA}{dt} = J_1 (V_2 - V_1) - \sum J (V_{+0} - V_{-0}).$$

The summation here for all of the surfaces of discontinuity that are found inside the piece of conductor is based on the product $J(V_{+0} - V_{-0})$.†

Section 64

Heating The Conductor. Joule's Law

In addition to the galvanic current, the electromotive forces produce another effect, namely, they heat the conductor. The mechanical theory of heat advances the theorem that the mechanical force of a system is the measure of the quantity of heat contained in it. We will designate P as the potential of all of the forces of attraction and repulsion that occur between the component parts of the system. Then, according to section 37, $\Sigma \frac{1}{2} m v^2 - P$ is the expression for mechanical force. If there are no other forces operating besides these forces of attraction and repulsion, then what one has, according to equation (2) in section 37, is

$$(1) \quad \sum \frac{1}{2} m v^2 - P = Const.$$

In this case, the system's mechanical force is unchangeable, and, therefore, the quantity of heat present in the system is also constant. If, however, other exterior forces occur, besides those interior forces of attraction

† Compare sections 57, 58, 59, 60, and 63 to *Kirchhoff,* über die Anwendbarkeit der Formeln für die Intensitäten der galvanischen Ströme in einem Systeme linearer Leiter auf Systeme, die zum Theil aus nicht linearen Leitern bestehen. *Poggendorff,* Annalen. Bd. 75. S. 189.

and repulsion whose components in point (x, y, z) may be denoted as X, Y, Z, then the theory of the conservation of kinetic energy will now read

$$(2)\quad \sum \frac{1}{2} m v^2 - P - \sum \int_0^t \left(X \frac{dx}{dt} + Y \frac{dy}{dt} + Z \frac{dz}{dt}\right) dt = Const.$$

And, therefore, in this case, the mechanical force of the system is

$$(3)\quad \sum \frac{1}{2} m v^2 - P = Const. + \sum \int_0^t \left(X \frac{dx}{dt} + Y \frac{dy}{dt} + Z \frac{dz}{dt}\right) dt,$$

and in the period from t=0 up to t, it is has increased by the work performed by those exterior forces during the same period. So the quantity of heat present in the system has increased by a quantum which is proportional to the work performed by the exterior forces, according to the theorem cited above of the theory of mechanical heat.

When this is applied to the previous case, what we have is

$$X = \varepsilon\left(\Xi + \frac{\partial V}{\partial x}\right),\quad Y = \varepsilon\left(\mathrm{H} + \frac{\partial V}{\partial y}\right),\quad Z = \varepsilon\left(\mathrm{Z} + \frac{\partial V}{\partial z}\right).$$

According to the section 59, the work performed by these forces in time element dt is

$$(4)\quad dA = dt \,.\int dS \,.\, w\, i^2,$$

and, consequently, the increase in heat that has occurred in this same time element dt is proportional to this magnitude.

In the special case, where the exterior electromotive forces $\epsilon\Xi$, ϵH, ϵZ only occur at points where any two heterogeneous parts of the conductor components touch, and where their result is constant for all points of the same surface of discontinuity, and is perpendicular to it, the transformations in the previous sections are valid for dA. What will then occur in the entire conductor during time element dt is an increase in heat, which is proportional to the work

$$(5)\quad -dt \sum J\,(V_{+0} - V_{-0}).$$

If what is being considered is only a section of the conductor, then what results is an increase in heat for this section in time dt which is proportional to the work

$$(6)\quad dt\,\{J_1(V_2 - V_1) - \sum J\,(V_{+0} - V_{-0})\}.$$

According to section 61, the work performed in time element dt for a wire-shaped section of the conductor can be expressed by $J^2 W\,dt$. If we assume J to be independent of t, then, in a unit of time, the conductor will receive an increase in heat which is proportional to the work

(7) $J^2\ W$

This law, which was established by Joule,† has been repeatedly proven experimentally.

† Philosophical Magazine. New and united series. Vol. XIX. 1841. pg. 260.

SIXTH DIVISION

Magnetism, Electromagnetism, And Electrodynamics

Section 65

The Basic Law Of Magnetic Interaction. The Potential Function Of Magnetic Forces

We can set up a hypothesis similar to that upon which the theory of electricity is based to explain magnetic phenomena. We will assume two opposing fluids to exist, a positive and a negative one. Two magnetic particles then, whose magnetic masses (according to numerical value and sign) are μ_1 and μ_2, will at distance r exert a force $\mu_1 \mu_2/r^2$ on each other in the direction of the connecting line between the particles. The force is either of repulsion or attraction, depending on whether the product of $\mu_1\mu_2/r^2$ is positive or negative. To the extent that attraction can be viewed as negative repulsion, one can also say: *two magnetic particles with magnetic masses of* μ_1 and μ_2 *which are at distance r from each other exert a repulsion of*

$$(1) \quad K = \frac{\mu_1 \mu_2}{r^2}$$

on each other in the direction of their connecting line.

The magnetic unit of mass here is that quantity of magnetic fluid which will repel a quantity similar to it in a unit of distance with a unit of force.

What we mean by a magnet is a ponderable body which contains magnetic fluid in such a distribution inside itself that it exerts magnetic effects towards the exterior. Experience teaches that no magnet can transfer any of the magnetic fluid contained inside it to the exterior, and that in every experimentally representable magnet the *algebraic sum of the magnetic masses is equal to zero:*

$$(2)\quad \sum \mu = 0, \quad \int d\mu = 0,$$

respectively, depending on whether the magnetic fluids are distributed in discrete points or continuously across the magnet.

We have excluded the data expressed in equations (2) from the effect that the earth as a magnet exerts on every experimentally representable magnet. The magnetic force of the earth can be broken down into a vertical and a horizontal component. If one then suspends a magnet so that it can move freely only in a horizontal direction, then the vertical component of the terrestrial magnetic force does not come into play. The horizontal component has a determinate magnitude and a determinate direction for every spot on the earth's surface. Therefore, due to the proportionally slight expansion of the suspended magnet, parallel and equal accelerations are imparted to all of its magnetic particles. If we denote this acceleration as T, then the total horizontal force acting on the magnet in the direction of the terrestrial magnetic meridian is $\Sigma T\mu = T\Sigma\mu$, $\int T d\mu = T\int d\mu$, respectively. But the terrestrial magnet does not exert any kind of attraction or repulsion, only a rotational effect. Consequently, it must be the case the $T\Sigma\mu = 0$ and $T\int d\mu = 0$, respectively; i.e., it must be that one or the other of the equations in (2) is satisfied because T is constant and different from zero.

The empirical theorems expressed above are satisfied through the assumption that in every molecule of the magnet there exist the same quantities of magnetic fluid which cannot cross over from one molecule to another under any circumstances. The magnet is said to be magnetized to its maximum when the magnetic fluid is distributed inside every molecule in such a way that the total effect towards the outside is a maximum.

The potential function originating in a magnet is:

$$(3)\quad V = -\sum \frac{\mu}{r}$$

or

$$(4)\quad V = -\int \frac{d\mu}{r},$$

depending on whether the fluid is assumed to be concentrated in discrete points or to be distributed continuously. Here, r will denote the distance of magnetic particles μ and $d\mu$, respectively, from point (x, y, z). The summation in (3) and the integration in (4) is to be extended over all the components of the magnet.

According to this, a force is operating on the aforesaid positive unit of magnetic mass that is concentrated in point (x, y, z). Its components parallel to the coordinate axes will be expressed by the equations

$$(5)\quad X = \frac{\partial V}{\partial x}, \quad Y = \frac{\partial V}{\partial y}, \quad Z = \frac{\partial V}{\partial z}.$$

Outside of the magnetic masses in which potential function V originates

$$\frac{\partial^2 V}{\partial x^2} + \frac{\partial^2 V}{\partial y^2} + \frac{\partial^2 V}{\partial z^2} = 0$$

everywhere, or, what says the same thing,

$$(6)\quad \frac{\partial X}{\partial x} + \frac{\partial Y}{\partial y} + \frac{\partial Z}{\partial z} = 0.$$

What furthermore results from equations (5) is that X, Y, Z must satisfy the partial differential equations

$$(7)\quad \frac{\partial Y}{\partial z} - \frac{\partial Z}{\partial y} = 0, \quad \frac{\partial Z}{\partial x} - \frac{\partial X}{\partial z} = 0, \quad \frac{\partial X}{\partial y} - \frac{\partial Y}{\partial x} = 0.$$

Although in reality we can only become acquainted with physical magnets, it is nevertheless not superfluous to consider the ideal case where the magnetic fluid is continuously distributed across a surface. The integration in (4) and (2) is then to be extended over all the elements of this surface.

Because equation (4) has the same form as equation (2) in section 45, the conclusions which were expressed there in equations (6) and (7) are also valid here. Thus, if one knows the potential function V that originates in a magnet for all of infinite space, then the magnetic density ϱ at point (x, y, z) is easily found. One obtains

$$(8)\quad \frac{\partial^2 V}{\partial x^2} + \frac{\partial^2 V}{\partial y^2} + \frac{\partial^2 V}{\partial z^2} = 4\pi\rho$$

when the magnetic masses are continuously distributed over a three-dimensional area, while, on the other hand, one obtains

$$(9)\quad \left(\frac{\partial V}{\partial p}\right)_{+0} - \left(\frac{\partial V}{\partial p}\right)_{-0} = 4\pi\rho$$

when they are continuously extended across a surface.

Section 66

The Magnetic Effects Of Galvanic Current

Experience shows that not only magnets but also galvanic currents exert magnetic effects towards the exterior. In order to investigate these effects, we will set up the ***hypothesis*** that the magnetic forces which are based in galvanic current are subject to the same laws everywhere *outside* of the current as they would have been had they originated in magnetic masses.

Let the galvanic current be linear and form a loop. A line (an infinitely thin wire) will be the conductor of the current, the line's endpoint will coincide with its beginning point, and it will have no component parts that either cut or overlap one another between the beginning and the endpoint. The positive unit of the magnet mass will be concentrated in point (x, y, z) which lies anywhere *outside* of the conductor. The galvanic current will exert a magnetic force on this positive unit whose components parallel to the coordinate axes will be denoted as X, Y, Z. According to the hypothesis that we have set up, these components satisfy the partial differential equations

$$(1)\quad \frac{\partial X}{\partial x}+\frac{\partial Y}{\partial y}+\frac{\partial Z}{\partial z}=0,\quad \frac{\partial Y}{\partial z}-\frac{\partial Z}{\partial y}=0,$$

$$(2)\quad \frac{\partial Z}{\partial x}-\frac{\partial X}{\partial z}=0,\quad \frac{\partial X}{\partial y}-\frac{\partial Y}{\partial x}=0.$$

As a consequence of equations (2) $X dx+Y dy+Z dz$ is a complete differential, and so there is a function V for x, y, z which is constituted in such a way that everywhere outside of the galvanic current

$$(3)\quad X=\frac{\partial V}{\partial x},\quad Y=\frac{\partial V}{\partial y},\quad Z=\frac{\partial V}{\partial z}.$$

Let us now extend the integral

$$(4)\quad \int(X\,dx+Y\,dy+Z\,dz)$$

by means of a curve traveling in a finite domain, whose endpoint coincides with its beginning point and whose remaining points can only be touched once in a simple circuit of the entire wire. In order to come to a clear understanding about this integral's value, it is desirable to set up a lemma in advance.

Section 67
Lemma From Analysis

We will draw a curve in the finite domain of the xy plane whose end-point and beginning point will coincide and whose remaining points can only be touched once in a simple circuit of the entire curve. The surface segment that will be enclosed by the curve [Figure 36] will totally lie in the quadrant in which x and y are positive. We define the positive sense of the curve by the fact that the tangent in the direction of the increasing arc will lie in the same relationship to the normal drawn towards the interior as the positive x-axis does in relation to the positive y-axis.

Let P and Q, then, be two functions of x and y which are single-valued, finite, and continuously variable inside the surface segment bounded by the curve. Let integral

$$(1) \quad -\int\int\left(\frac{\partial P}{\partial y}-\frac{\partial Q}{\partial x}\right)\delta x\,\delta y$$

be extended over the surface segment bounded by the curve. δx and δy will denote *positive* increases in the variables. We can begin with integration for y for the first component of the integral, and we will draw the ordinates which belong to the abscissas x and $x+\delta x$. Between them lies an infinitely narrow strip which enters the surface domain bounded by the curve just as often as it leaves the same. We will describe the ordinates at the place of entry as $y_1, y_3, y_5, \ldots y_{2n-1}$ while, on the other hand, we will denote the or-

Figure 36

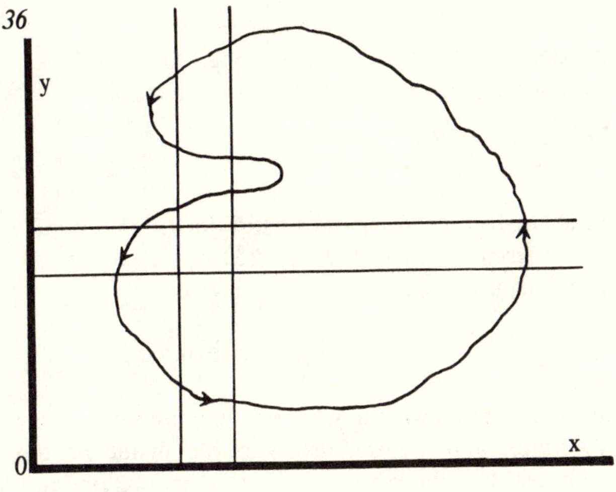

dinates at the place of exit as $y_2, y_4, y_6, \ldots y_{2n}$ and we notice that $y_1 < y_2 < y_3 < \ldots y_{2n-1} < y_{2n}$. The arc elements that the infinitely narrow strip will cut off in its entry and exit from the boundary curve are $ds_1, ds_2, ds_3, \ldots ds_{2n-1}, ds_{2n}$. The cosine of the angle that such an arc element will form with the direction of the positive x is positive at all points of entry, and negative at all the points of exit. If one thus denotes (according to numerical value and sign) dx_k as the projection of ds_k on the x-axis, then what results is $dx_1 = dx_3 = dx_5 = \ldots = dx_{2n-1} = \delta x$, $dx_2 = dx_4 = dx_6 = \ldots = dx_{2n} = -\delta x$. According to this we will find that

$$\delta x \int \frac{\partial P}{\partial y} \delta y = \delta x (-P_1 + P_2 - P_3 + \ldots + P_{2n})$$

$$= - P_1 \, dx_1 - P_2 \, dx_2 - \ldots - P_{2n} \, dx_{2n},$$

which can be abbreviated as

$$\delta x \int \frac{\partial P}{\partial y} \delta y = -\sum P \, dx,$$

in which the sum sign on the right-hand side means that the value for Pdx should be taken for all the entry and exit places of the infinitely narrow surface strip. The integration will be carried out for x so that one not only takes into consideration a single strip, but all of them that (bounded by the parallels to the y-axis) generally cut through the surface domain. Consequently, the result is

$$\int\int \frac{\partial P}{\partial y} \delta x \, \delta y = - \int P \, dx$$

and the integration on the right is to be extended through the entire loop.

In a corresponding manner, we will come to equation

$$\int\int \frac{\partial Q}{\partial x} \delta y \, \delta x = \int Q \, dy$$

and here too the integration on the right-hand side (in the circuit's positive direction) is to be extended through the entire loop.

Our result will therefore be

$$(2) \quad - \int\int \left(\frac{\partial P}{\partial y} - \frac{\partial Q}{\partial x} \right) \delta x \, \delta y = \int (P dx + Q dy)$$

provided that we understand P and Q as two functions of x and y that are single-valued, finite, and continuously variable inside of the surface domain of the integral on the left-hand side of the equation, and that one extends the integration on the right in the circuit's positive direction along the boundary curve.†

In the special case where

$$(3)\quad \frac{\partial P}{\partial y} - \frac{\partial Q}{\partial x} = 0$$

everywhere inside of the surface domain enclosed by the curve, then equation (2) changes into

$$(4)\quad \int (P\,dx + Q\,dy) = 0.$$

Moreover, as one can easily see, theorems (2) and (4) also remain valid when the curve does *not* entirely run its course in the positive xy quadrant. This assumption only serves to make the development of the proof easier. If this assumption is not satisfied from the outset, then we can easily establish that the desired situation exists through parallel displacement of the axes, because the curve does not extend into the infinite anywhere.

Section 68

The Integral $\int (Xdx+Ydy+Zdz)$.

After this preparation, we can now return to integral (4) in section 66. We will use a continuously curved surface, all of which lies in the finite domain, and which has the prescribed path of integration as its total and exclusive boundary. Let the equation of this surface be

$$(1)\quad z = f(x, y).$$

Since the prescribed path of integration lies on the surface, then, in section 66, equation (4), one has to put

$$dz = \frac{\partial f}{\partial x}\,dx + \frac{\partial f}{\partial y}\,dy$$

and eliminate coordinate z from X, Y, Z, using equation (1). Thereby, integral (4) in section 66 will change into

$$(2)\quad \int \left\{ \left(X + Z\frac{\partial f}{\partial x}\right) dx + \left(Y + Z\frac{\partial f}{\partial y}\right) dy \right\},$$

† *Riemann.* Grundlagen für eine allgemeine Theorie der Functionen einer veränderlichen complexen Grösse. Göttingen 1851. Art. 7 und 8.

and the integration here is to be extended by means of the projection of the given curve lying in the x y plane.

The coordinate system can always be placed in such a way that this projection will loop just as simply as the projected curve itself, and that, with integration in (2), this projection will travel in a positive direction, that is, (viewed from a point on the positive x-axis) the tangent in the direction of the increasing arc will lie in the same relationship to the normal which is drawn towards the interior as the positive x-axis lies in relationship to the positive y-axis. [Figures 36 and 37] In this situation, which we are allowed to presume on account of uniformity and without detriment to the universality of the investigation, surface (1) can also be arranged so that its projection is *simply* a segment of the x y plane which the projection of its boundary curve encloses. Then function z, which is expressed by equation (1), along with its first derivatives is single-valued, finite, and continuously variable for the entire domain of values for x and y concerned. Because functions X, Y, Z are also single-valued, finite, and continuously variable, we can then specifically have

$$(3)\qquad P = X + Z\frac{\partial f}{\partial x},\quad Q = Y + Z\frac{\partial f}{\partial y},$$

in our investigation in section 67. What results from this is

$$\frac{\partial P}{\partial y} = \frac{\partial X}{\partial y} + \frac{\partial X}{\partial z}\frac{\partial f}{\partial y} + \left(\frac{\partial Z}{\partial y} + \frac{\partial Z}{\partial z}\frac{\partial f}{\partial y}\right)\frac{\partial f}{\partial x} + Z\frac{\partial^2 f}{\partial x\,\partial y},$$

$$\frac{\partial Q}{\partial x} = \frac{\partial Y}{\partial x} + \frac{\partial Y}{\partial z}\frac{\partial f}{\partial x} + \left(\frac{\partial Z}{\partial x} + \frac{\partial Z}{\partial z}\frac{\partial f}{\partial x}\right)\frac{\partial f}{\partial y} + Z\frac{\partial^2 f}{\partial x\,\partial y}.$$

Figure 37

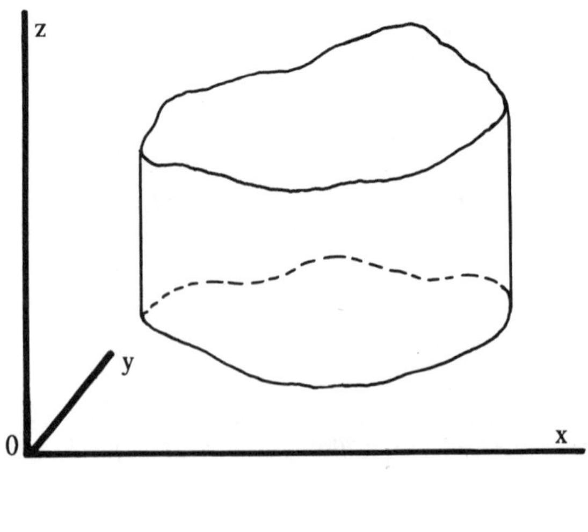

And, consequently, we have

$$\frac{\partial P}{\partial y}-\frac{\partial Q}{\partial x}=\left(\frac{\partial X}{\partial y}-\frac{\partial \dot{Y}}{\partial x}\right)-\left(\frac{\partial Y}{\partial z}-\frac{\partial Z}{\partial y}\right)\frac{\partial f}{\partial x}$$

(4)

$$-\left(\frac{\partial Z}{\partial x}-\frac{\partial X}{\partial z}\right)\frac{\partial f}{\partial y}.$$

Now, if the curve by means of which integral $\int(X\,dx+Y\,dy+Z\,dz)$ is supposed to extend is not entangled like a chain *around the conductor of the galvanic current,* then surface (1) can always be placed in such a way that it does not have any point in common with this conductor. Then equations (2) in section 66 will be satisfied for every point in the surface, and, consequently, equation (4) will change into

(5) $$\frac{\partial P}{\partial y}-\frac{\partial Q}{\partial x}=0.$$

This is also valid for equation (4) in section 67, i.e.,

(6) $$\int(X\,dx+Y\,dy+Z\,dz)=0,$$

when this integral is extended along a loop which is not entangled like a chain around the conductor of the galvanic current.

But if, on the contrary, the integration curve is entangled around the conductor in the form of a chain, then it is impossible to place surface (1) so that it does not have any point in common with the conductor. Equation (5) is then not satisfied everywhere, and integral (6) does not have a value of zero.

Section 69

The Potential Function V Of Electromagnetic Forces

We will set up surface S in such a way that the conductor of the galvanic current forms its total and sole boundary. Then theorem (6) in section 68 will be valid for every path of integration that does not cut surface S. If we assume that any place in space is the integration's beginning point and then extend integral

(1) $$\int(X\,dx+Y\,dy+Z\,dz)$$

out from there to point (x, y, z) by *different* paths, none of which, however, cut surface S, the integral values that occur on all of these paths will be equal to each other. So integral (1) is a function V of x, y, z that changes continuously everywhere outside of surface S.

We will now set up a normal at any point in surface S going toward both sides and count off the interval p on it from its base, positive on one side and negative on the other. We then take one point each on the positive and the negative normal that are infinitely close to surface S and denote the values which function V assumes in them as V_{+0} and V_{-0} respectively. The difference $V_{+0}-V_{-0}$ will result when one extends integral (1) from a point on the negative normal to the point lying infinitely close to surface S on the positive normal by means of a curve that does not cut surface S.

X, Y, and Z, however, are single-valued, finite, and continuously variable in all of infinite space outside of the current's path. It follows from this that at infinite proximity to surface S, V's derivatives, taken parallel as well as perpendicular to the surface, have the same value on the positive side as on the negative side. So, for every arbitrary place on the surface,

$$(2) \quad V_{+0} - V_{-0} = A,$$

where A denotes a constant magnitude and, furthermore,

$$(3) \quad \left(\frac{\partial V}{\partial p}\right)_{+0} = \left(\frac{\partial V}{\partial p}\right)_{-0}.$$

According to equation (1) in section 66, function V satisfies the partial differential equation

$$(4) \quad \frac{\partial^2 V}{\partial x^2} + \frac{\partial^2 V}{\partial y^2} + \frac{\partial^2 V}{\partial z^2} = 0,$$

for all of infinite space outside of the current's path, and it will have a value of zero at an infinite distance:

$$(5) \quad V = 0 \text{ for } x^2 + y^2 + z^2 = \infty.$$

So, according to these conditions, (2) through (5), function V only and always comes about in one way, which we can easily prove with the aid of *Dirichlet's* Principle.

But instead of this, we will form the expression for function V directly according to *Green's* method.

Section 70

Establishing Function V

In this, *Green's* theorem (section 20) can be used as follows:
We put

$$\sqrt{(x-x')^2+(y-y')^2+(z-z')^2}=r$$

and

$$(1)\quad U=\frac{1}{r}.$$

Space T, which is what we will be taking into consideration, is bounded by both sides of surface S and by two spherical surfaces of which one will have the radius r=R and the other will have the radius r=c, with point (x′, y′, z′) as the center, where lim R=∞ and lim c=0.

All the assumptions in section 21 will hold here, with the one modification that U is not zero everywhere on the surface of space T. By repeating the procedure that we used there, we will get the equation

$$(2)\quad \begin{aligned} 4\pi V' = & -\int\int\int U\left(\frac{\partial^2 V}{\partial x^2}+\frac{\partial^2 V}{\partial y^2}+\frac{\partial^2 V}{\partial z^2}\right) dx\, dy\, dz \\ & +\int d\sigma\left(V\frac{\partial U}{\partial n}-U\frac{\partial V}{\partial n}\right), \end{aligned}$$

and V′ denotes the value for V in point (x′, y′, z′). According to equation (4) in section 69, the volume integral is zero. The surface integral is to be extended over the sphere with a radius of R and over both sides of surface S.

For the spherical surfaces, the normal drawn towards the interior of space T will fall in the direction of decreasing r. Consequently, the integral that extends over the spherical surface can be written as

$$-R^2\int d\omega\left(V\frac{\partial U}{\partial r}-U\frac{\partial V}{\partial r}\right)_{r=R},$$

where dw designates an element of the spherical surface of radius one. Now, for lim R=∞, V as well as U have a finite ratio to 1/R. Consequently,

$$\lim\left(V\frac{\partial U}{\partial r}-U\frac{\partial V}{\partial r}\right)_{r=\infty}=\frac{const.}{R^3},$$

and one can see that the integral has the limit zero for lim R=∞.

Hereafter, nothing else remains on the right-hand side of equation (2) except the surface integral, extended over both sides of surface S.

We will now construct a normal at point (x, y, z) of surface S going towards both sides and will count off on it the distance p starting from its base, positive on one side and negative on the other. Then n=p on the positive side and n= -p on the negative side. Equation (2) will therefore change into the following:

$$4\pi V' = \int d\sigma \left\{ V\frac{\partial U}{\partial p} - U\frac{\partial V}{\partial p} \right\}_{p=+0} - \int d\sigma \left\{ V\frac{\partial U}{\partial p} - U\frac{\partial V}{\partial p} \right\}_{p=-0},$$

which one can also write as

$$4\pi V' = \int d\sigma \,.\, \frac{\partial U}{\partial p}\,(V_{+0} - V_{-0}) - \int d\sigma \,.\, U\left\{\left(\frac{\partial V}{\partial p}\right)_{+0} - \left(\frac{\partial V}{\partial p}\right)_{-0}\right\}.$$

The second integral is equal to zero here as a result of equation (3) in section 69. In the first integral, one has to take into account equation (2) in section 69 for V, and equation (1) in this section for U. Therefore, what finally results is

$$(3) \qquad 4\pi V' = A\int d\sigma \,.\, \frac{\partial\left(\frac{1}{r}\right)}{\partial p}.$$

Section 71

Continuation

We will further advance the hypothesis that the magnetic forces, which are exerted by *several* galvanic currents on the positive magnetic unit of mass, assumed concentrated in point (x, y, z), are composed according to the theorem of the parallelogram of forces. What directly follows from this for any single current is that an n-fold intensity of current also exerts n-fold forces. Therefore, components X, Y, Z are proportional to current intensity J, and so

$$A = k\,J.$$

Because we are always taking the current intensity in the direction of increasing arc length s of the current conductor (section 61), what is to be determined first of all is how this direction and the direction of the positive normal constructed on surface S should lie in relation to each other. We find that, viewed from a point on the positive normal constructed on S, the direction of the increasing arc s will be the same as the clockwise direction. Or in other words: If anyone places himself upright on the positive side of surface S and then goes along the boundary in the direction of increasing arc s, surface S will be on his *right.*

Experience teaches us that, given this relationship of the positive normal to the increasing arc, the difference $V_{+0}-V_{-0}$ turns out to be positive whenever J is positive. Consequently, constant k is positive. For the sake of simplicity, we shall select the unit of current intensity so that $k=4\pi$. Equation (2) in section 69 will then change into

$$(2) \quad V_{+0} - V_{-0} = 4\pi J,$$

and instead of equation (3) in section 70, we obtain

$$(3) \quad V' = J\int d\sigma \frac{\partial\left(\frac{1}{r}\right)}{\partial p}.$$

The mass of the current intensity introduced here will be called the *magnetic* mass.

Section 72

The Mechanical Significance Of The Expression For V

We will investigate the mechanical significance of the expression found for V′. It is

$$\frac{\partial\left(\frac{1}{r}\right)}{\partial p} = \lim \frac{\left(\frac{1}{r}\right)_{p=\delta} - \left(\frac{1}{r}\right)_{p=0}}{\delta}$$

for lim $\delta=0$. Consequently, one can write V in such a way that

$$(1) \quad V' = \lim \int d\sigma \left\{ \frac{J}{\delta}\left(\frac{1}{r}\right)_{p=\delta} - \frac{J}{\delta}\left(\frac{1}{r}\right)_{p=0} \right\}.$$

The expression for V′ then has the form of the potential function of an ideal magnetic distribution of mass. If we imagine that one can continuously expand the magnetic fluids over a surface and that $d\mu$ is the magnetic mass contained in surface element $d\sigma$, then

$$V = -\int \frac{d\mu}{r}$$

would be the potential function for this magnetic mass, if one extends the integral across the entire surface.

If we then cover every surface element of surface S (for p=0) with the magnetic mass

$$d\mu = \frac{J}{\delta} d\sigma$$

and cover every surface element in a surface parallel to S (for $p=\delta$) with the magnetic mass

$$d\mu = -\frac{J}{\delta} d\sigma,$$

then the effect of the magnetic masses extended across both surfaces will be the same as the effect of the galvanic current passing through S's boundaries.

Both magnetic masses will then have opposite signs and an equal and constant density, which is inversely proportional to the distance of the separation.

This does not include the effect on a point in interior space which is between the surfaces which lie infinitely close to each other.

Section 73

The Geometric Significance Of The Expression For V

The integral in equation (3) in section 71 is also *geometrically significant.* We will draw a ray from point (x', y', z') which will bisect current path s and we will set the ray in motion in such a way that the intersection point will travel along curve s from beginning to end. This will produce a conical surface that has point (x', y', z') as its apex, and we will set up a spherical surface with the radius of one around that same point as its midpoint. This spherical surface will be bisected by the conical surface along a closed line s_1. For the sake of simplicity, we will first assume that line s_1 (the projection of s on the sphere) will be looped just as line s itself is, so that when one travels along it from beginning to end, one will not hit any of its points more than once. We have not made any kind of special provision at all about the shape of surface S that is bounded by current path s. For just as we agreed above, it is only when the projection of s_1 on a sphere with a radius of one is a line that is looped can we give that kind of a shape to surface S so that its projection will simply cover a piece of the spherical surface that is *enclosed* by s_1. If one were then to draw a ray from (x', y', z') through any point on this enclosed piece of the spherical surface, then, if the ray were properly extended, it would bisect surface S at a point, in one and *only* one point. At this place, the extended ray would cross over from the side of surface S *facing towards* point (x', y', z') to the side *facing away from* this point. The part of the spherical surface that is not covered by the projection of surface S can then be designated as the domain *excluded* by s_1. If one then draws a ray from (x', y', z') through any point on this excluded domain, then no matter how far one may extend it, this ray does not meet surface S.

We will select any point (x, y, z) on surface S and denote r as the length of the ray drawn from point (x', y', z') to (x, y, z). Let $d\sigma$ denote a surface element of surface S on whose boundary point (x, y, z) lies. If we now allow a moving ray starting from (x', y', z') to glide around $d\sigma$'s entire boundary, then the ray will portray a conical surface. This conical surface will cut the spherical surface with a radius of one along a simple loop, which is the boundary of surface element $d\Sigma$ lying on the sphere and surface element $d\Sigma$ is the projection of $d\sigma$.

Positive normal p, which has been set up at point (x, y, z) against surface S, will form an angle along with the direction of the increasing r whose cosine will be

$$\cos(r\,p) = \frac{\partial r}{\partial p}$$

and this cosine is either positive or negative, depending on whether the side of surface S that is *turned away* from point (x', y', z') is either positive or negative.

The calculations for a projection of $d\sigma$ onto a sphere inscribed around (x', y', z') with a radius of r are

$$= \pm\, d\sigma \frac{\partial r}{\partial p};$$

and it will be valid for either the plus or the minus sign depending on whether $\partial r/\partial p$ is positive or negative. If projection $d\Sigma$ is to be expressed on a sphere with a radius of one, then one still has to divide by r^2, so that

$$(1)\quad d\Sigma = \pm \frac{1}{r^2}\frac{\partial r}{\partial p}\, d\sigma = \mp\, d\sigma \frac{\partial\left(\frac{1}{r}\right)}{\partial p}.$$

We will denote Σ as the surface area of the projection of S on a sphere with a radius of one, with the understanding that Σ is an absolute number. Then, what results from (1) through integration is

$$(2)\quad \int d\sigma \frac{\partial\left(\frac{1}{r}\right)}{\partial p} = \mp \Sigma,$$

and it will be positive or negative depending on whether surface S's positive or negative side is *turned towards* point (x′, y′, z′).

But attention still must be paid to the signs for J. If surface S's positive side is turned towards point (x′, y′, z′), then J is either positive or negative depending on whether — looking outwards from the point — the positive current flows clockwise or counterclockwise. If surface S faces its negative side toward point (x′, y′, z′), then the rule for J's sign is reversed. This can also be expressed in another way, namely: J's sign will be the same as the sign on the right-hand side of equation (2) when — looking outwards from point (x′, y′, z′) — the direction of the positive current agrees with the direction of rotation of the hands of a clock. And J's sign will be the opposite of the sign on the right-hand side of equation (2) when — looking outwards from point (x′, y′, z′) — the direction of the positive current is the opposite of a clock's direction of rotation. So product

$$J \int \frac{\partial\left(\frac{1}{r}\right)}{\partial p} d\sigma$$

is either positive or negative depending on whether — viewed from point (x′, y′, z′) — the positive or the negative current flows in the direction of rotation of the hands of a clock.

Figure 38

For an observer situated at point (x′, y′, z′), what we mean by the term *celestial sphere* is the sphere constructed around this point as its center with a radius of one.

So hereafter, one has the following rule for establishing the potential function in point (x′, y′, z′):

One multiplies the absolute value of the current intensity by that section of the celestial sphere which appears to be enclosed by the current's path for an observer situated in point (x′, y′, z′). *One then gives this product either a positive or negative sign depending on whether, for that same observer, the direction of the positive current agrees with the direction of rotation of the hands of a clock or is opposite to it.*

This rule, stated by *Gauss*,† remains valid when the

† Allgemeine Theorie des Erdmagnetismus, Art. 38. (Resultate aus den Beobachtungen des magnetischen Vereins, 1838. — *Gauss's* Werke, Bd. 3.)

projection of the simple, self-enclosed current path on the celestial sphere includes a double point. Then surface S will no longer just be turning one side toward point (x′, y′, z′). Rather [Figure 38] those two components of Σ whose boundaries meet at a double point will belong to those sections of surface S that turn opposite sides toward point (x′, y′, z′). So according to equation (2), those components of Σ which follow each other in a single file are all calculated with alternating signs and they will all have the same J and the same kind of sign as their common factor. But, if one wants to calculate the absolute value for one single component for Σ by multiplying it by the absolute value for J, then one has to give alternating signs to the products and so one returns to the rule set up by *Gauss*.

Section 74

The Effect Of A Single Element Of Current On A Single Magnetic Particle

Since we have now succeeded in establishing a geometric interpretation for formula (3) in section 71 that expresses the potential function of the magnetic force exerted by the galvanic current, we can apply this to the components of force themselves. In order to have component X parallel to the positive x-axis, we have to displace point (x′, y′, z′) by a distance of δx in this direction and divide the change proceeding in function V′ through this by the magnitude of the displacement. Function V′ is now the product of two factors, of which the first — current intensity J — does not undergo any change due to the displacement that was undertaken. The second factor is that part of the celestial sphere which an observer sees as enclosed by the current's path at one time from point (x′, y′, z′), and then at the next time from point (x′+δx, y′, z′). But, instead of having the current's path fixed in space and displacing the observer's viewpoint from point (x′, y′, z′) to point (x′+δx, y′, z′), one can also have the observer remain in the first point and displace the current's path in relationship to it in such a way that every one of its points will travel along δx's path in the direction of decreasing x. The question, then, is which segment of the celestial sphere leaves the enclosure as a result of this, and which new segments enter it. [Figures 39 and 40] As a result of the displacement, every element ds in the current's path creates an infinitely small parallelogram whose projection can be found on the celestial sphere. We will permit the direction of the increasing arc to coincide with the direction of the positive current. In Figure 39, this direction will agree with the direction of rotation of a clock's hand, while in Figure 40 the direction will be the opposite. If one calculates V′ by multiplying the absolute values of the projection on the celestial sphere and the

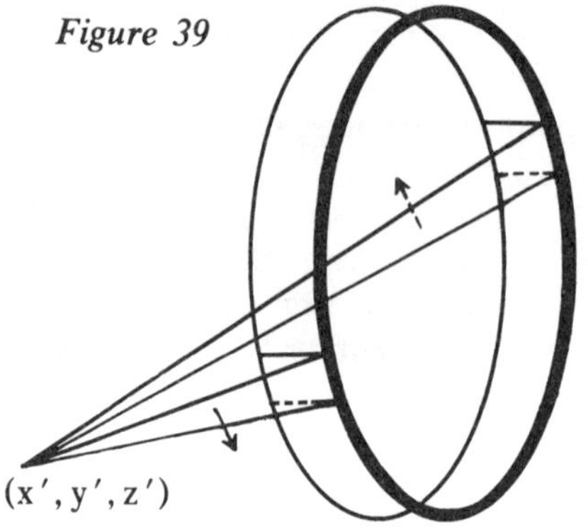

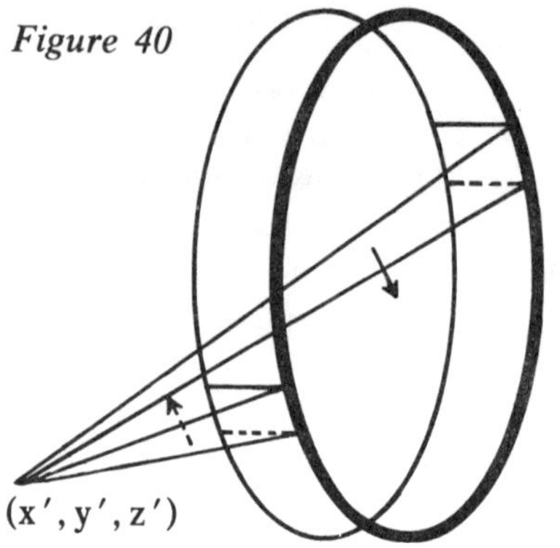

current intensity by each other, then a positive sign is to be given to the product for Figure 39, and a negative one given to the product for Figure 40. Therefore, one will obtain the change in V′ with the correct signs if, in Figure 39, one calculates those segments of the celestial sphere that are leaving the enclosure with minus signs, those that are entering the enclosure with plus signs, and then multiplies by the absolute value of the current intensity. But, in Figure 40, on the other hand, one has to take the part leaving the celestial sphere as positive, while those entering will be negative and then multiply here too by the absolute value of the current intensity.

If we now view arc element ds as an infinitely small straight line, then this line and point (x′, y′, z′) determine a plane. We will draw a normal p to this plane and, naturally, it will be positive on that side of the plane on which an observer who is standing upright and faces the current sees the positive current flowing by from the right to the left. We will denote the cosine of the angle that is formed by the direction of the positive normal and the direction of positive x as cos(x, p). What is then to be noted is that this cosine is either positive or negative for Figure 39, depending on whether arc element ds belongs to a parallelogram that is entering or exiting the enclosure. For Figure 40, the opposite rule holds.

Such a parallelogram has sides ds and δx, whose projections on the celestial sphere will be, respectively,

$$\frac{ds}{r} \sin (r, ds)$$

and

$$\pm \frac{\delta x}{r} \cos (x, p).$$

These projections are perpendicular to each other, for one lies in the plane of angle (r, ds) and the other lies in the normal of this plane. The parallelogram will be projected on the celestial sphere in the form of a rectangle, whose area is (disregarding signs)

$$= \frac{ds}{r} \frac{\delta x}{r} \sin (r, ds) \cos (x, p).$$

This product will have the same sign as cos(x,p), and, therefore, the product is either positive or negative for Figure 39, depending on whether the surface element that is expressed by this product enters or leaves the enclosure. The opposite signs will result with Figure 40. So, in every case, one will be able to calculate the element in question correctly if one takes the product itself along with its proper signs. Consequently, we get

(1) $$\delta V' = \delta x \, . \, J \int \frac{ds}{r^2} \sin (r, ds) \cos (x, p)$$

and

(2) $$X = J \int \frac{ds}{r^2} \sin (r, ds) \cos (x, p).$$

Here, δx and J are assumed to be absolute. We can take out of equation (2) that component of force which a single current element ds uses on the positive unit of magnetic mass concentrated in point (x′, y′, z′) in the direction of the increasing x. This component of force in the direction of the increasing x is

$$J \frac{ds}{r^2} \sin (r, ds) \cos (x, p).$$

The components of force in the direction of the increasing y and increasing z, respectively, are found in the same way:

$$J \frac{ds}{r^2} \sin (r, ds) \cos (y, p), \quad J \frac{ds}{r^2} \sin (r, ds) \cos (z, p).$$

Thus it follows that the total force that current element ds exerts on point (x′, y′, z′) falls in p's direction, i.e., in the direction of the *positive* normal to the plane, which is determined by current element ds and point (x′, y′, z′). This force's magnitude is

(3) $$J \frac{ds}{r^2} \sin (r, ds).$$

The positive normal leaves the plane in that very space in which a person standing upright on the plane will see the positive current flow from his right to his left.

This rule puts us in the position of being able to specify, according to magnitude and direction, those forces which are exerted by all of the current elements of a closed galvanic current or, also, by several currents on the unit of positive magnetic mass found at point (x', y', z'). The single forces are composed according to the law of parallelograms.

Section 75

Integral $\int(Xdx+Ydy+Zdz)$ And The Current Intensity

We will consider, as we did before, a closed linear current and we will set up a surface S so that the self-enclosing path of the current will form its single and complete boundary. [Figure 41] We will take the positive side of the surface to be that surface which must appear to be facing towards the observer so that he sees the positive current flowing clockwise. In terms of *magnetic* measurement this equation is valid for current intensity.

$$(1) \quad 4\pi J = \int dV = \int (X\,dx + Y\,dy + Z\,dz) = V_{+0} - V_{-0}$$

Figure 41

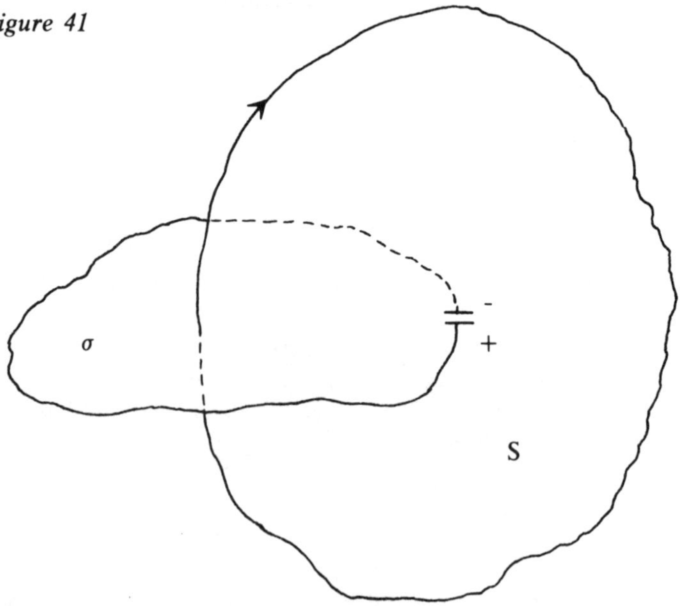

when the integral is extended through a curve that, without bisecting surface S, goes from a point on the negative side of the surface to the point lying infinitely close to it on the positive side. We will set up a surface σ in such a way that this path of integration constitutes its sole boundary. Surface σ will be bisected by the current's path. We will distinguish the positive and negative side of σ by having the positive current flow from the negative side through the surface to the positive side. So, when one places himself squarely on the positive side of σ and then travels along the path of integration in the manner followed by the integral in (1), he will have surface σ on his left hand. Or, in other words, the path of integration will go from right to left for an observer who stands upright on the positive side of surface σ.

If J is independent of time, then it will give the algebraic sum of the quantities of electricity which change over from the negative to the positive side of σ in the unit of time, reduced by the algebraic sum of those quantities which go over from the positive to the negative side in the same amount of time. One can calculate this surplus according to equation (1) by extending integral

$$(2) \quad \frac{1}{4\pi}\int (X\,dx + Y\,dy + Z\,dz)$$

through the boundary of σ in the prescribed direction. This also holds when surface σ is not bisected by the current's path, for in such cases, integral (2), as well as the quantity of electricity going through σ is equal to zero.

If there are various galvanic currents present that have these components of magnetic force —

$$X_1,\ Y_1,\ Z_1,$$
$$X_2,\ Y_2,\ Z_2,$$
$$\dots\dots\dots$$

then, according to the law of parallelograms, they can be written as

$$(3) \quad \begin{aligned} X &= X_1 + X_2 + \dots \\ Y &= Y_1 + Y_2 + \dots \\ Z &= Z_1 + Z_2 + \dots \end{aligned}$$

and this theorem will be generally valid:

The integral

$$\frac{1}{4\pi}\int (X\,dx + Y\,dy + Z\,dz),$$

which extends from left to right through σ's boundary, specifies how much larger that quantity of electricity is that goes over from the negative side to the positive side of σ in the unit of time, compared to the quantity of electricity which goes over in the opposite direction through surface σ during the same time.

Section 76

Specific Current Intensities Expressed By The Components Of The Electromagnetic Force

We now want to apply the theorem in section 75 to infinitely small surface elements. We first consider a plane rectangle whose sides dx and dy, respectively, run parallel to the x and y axes. The corner point lying closest to the origin will have coordinates x, y, z. The plane of the rectangle will lie perpendicular to the z-axis. The specific current intensity in the direction of this axis is i_3. Consequently, equation (1) in section 75 will change into the following here:

$$4\pi\, i_3\, dx\, dy = \int (X\, dx + Y\, dy + Z\, dz).$$

The integral on the right-hand side is to be extended in a positive direction along the rectangle's boundary. We will denote the rectangle's sides [Figure 42] in such a way that they will follow each other in a positive succession of 1, 2, 3, 4 and we will stipulate that side 1 leads from point (x, y, z) to point (x+dx, y, z). Furthermore, we will give components X, Y, Z, indices of 1, 2, 3, 4 in order to indicate that they represent the value of the side indexed with the same number. The integral, then, results in

$$X_1\, dx + Y_2\, dy - X_3\, dx - Y_4\, dy,$$

when taken throughout the rectangle's boundary. But

$$X_1 = X, \quad X_3 = X + \frac{\partial X}{\partial y}\, dy, \quad Y_4 = Y, \quad Y_2 = Y + \frac{\partial Y}{\partial x}\, dx.$$

Consequently, the integral will change into

$$\left(\frac{\partial Y}{\partial x} - \frac{\partial X}{\partial y}\right) dx\, dy$$

and equation (1) in section 75 will now read

Figure 42

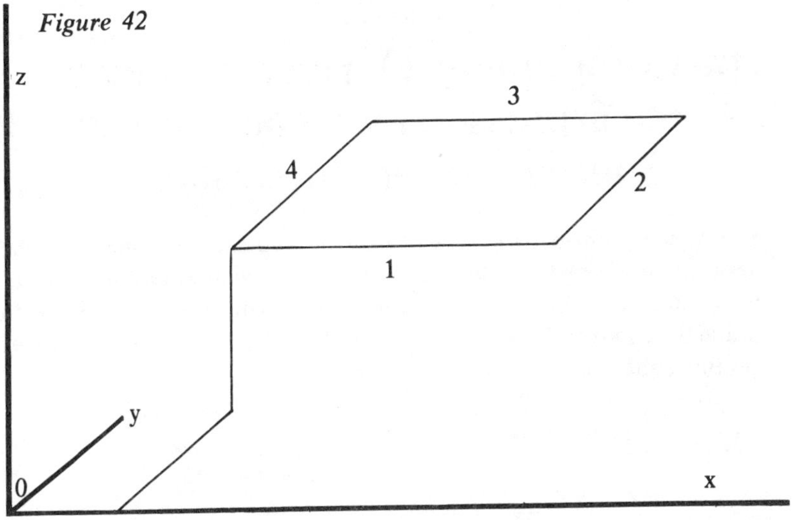

$$4\pi\, i_3\, dx\, dy = \left(\frac{\partial Y}{\partial x} - \frac{\partial X}{\partial y}\right) dx\, dy.$$

One obtains two corresponding equations when one sets up two more two-dimensional surface elements perpendicular to the x and y axes going through point (x, y, z). The results read

$$(1)\quad 4\pi\, i_1 = \frac{\partial Z}{\partial y} - \frac{\partial Y}{\partial z},$$

$$(2)\quad 4\pi\, i_2 = \frac{\partial X}{\partial z} - \frac{\partial Z}{\partial x},$$

$$(3)\quad 4\pi\, i_3 = \frac{\partial Y}{\partial x} - \frac{\partial X}{\partial y}.$$

These equations can serve to calculate the specific current intensities for place (x, y, z) when the components of the magnetic force exerted by the galvanic currents are given.

Section 77

The Components Of Electromagnetic Force Expressed By Means Of The Specific Current Intensities

Conversely, consider the task of determining the components of the magnetic force exerted by galvanic currents when the specific current intensities are given for every place in space. The question here is to determine functions X, Y, Z in such a way that they will satisfy the partial differential equations

$$(1)\quad \frac{\partial Z}{\partial y} - \frac{\partial Y}{\partial z} = 4\pi i_1,$$

$$(2)\quad \frac{\partial X}{\partial z} - \frac{\partial Z}{\partial x} = 4\pi i_2,$$

$$(3)\quad \frac{\partial Y}{\partial x} - \frac{\partial X}{\partial y} = 4\pi i_3.$$

The specific current intensities i_1, i_2, i_3 will only be different from zero inside of the conductor through which current flows and they will be equal to zero in all other space. Because we have provided that the magnetic forces will only originate in galvanic currents and not in magnetic masses, the partial differential equation (6) in section 65 is satisfied for all of infinite space, namely

$$(4)\quad \frac{\partial X}{\partial x} + \frac{\partial Y}{\partial y} + \frac{\partial Z}{\partial z} = 0.$$

It should be noted that at an infinite distance the magnetic forces are equal to zero:

$$(5)\quad X_\infty = Y_\infty = Z_\infty = 0.$$

Equations (1), (2), and (3) are not totally independent from each other. They will satisfy the condition

$$\frac{\partial i_1}{\partial x} + \frac{\partial i_2}{\partial y} + \frac{\partial i_3}{\partial z} = 0.$$

We will first of all eliminate Y and Z in order to solve our problem. This will happen when we differentiate by x in equation (4), by $-y$ in equation (3), by z in equation (2), and add the results left and right. What results is

$$\frac{\partial^2 X}{\partial x^2} + \frac{\partial^2 X}{\partial y^2} + \frac{\partial^2 X}{\partial z^2} = 4\pi\left(\frac{\partial i_2}{\partial z} - \frac{\partial i_3}{\partial y}\right).$$

By comparison with equation (4) in section 13, one can get a mechanical interpretation of the result. Hereafter, one can consider X as the potential function originating in an attracting gravitational mass when the density of the mass is

$$= -\left(\frac{\partial i_2}{\partial z} - \frac{\partial i_3}{\partial y}\right)$$

in point (x, y, z). What immediately results is

$$(6) \quad X' = -\int\left(\frac{\partial i_2}{\partial z} - \frac{\partial i_3}{\partial y}\right)\frac{dT}{r},$$

and in a corresponding manner

$$(7) \quad Y' = -\int\left(\frac{\partial i_3}{\partial x} - \frac{\partial i_1}{\partial z}\right)\frac{dT}{r},$$

$$(8) \quad Z' = -\int\left(\frac{\partial i_1}{\partial y} - \frac{\partial i_2}{\partial x}\right)\frac{dT}{r}.$$

In these equations, dT signifies the spatial element contiguous to point (x, y, z), r is the distance of point (x′, y′, z′) from point (x, y, z), and X′, Y′, Z′ are the components of the magnetic forces which are exerted on the positive unit of magnetic mass concentrated in point (x′, y′, z′). The integrations in equations (6), (7), and (8) are to be extended across all of the conductors that have galvanic current flowing through them.

Section 78

Continuation: Other Solutions To The Problem

We can also solve the problem in section 77 in another way. We will have

$$X = \frac{\partial u_2}{\partial z} - \frac{\partial u_3}{\partial y},$$
$$(1) \quad Y = \frac{\partial u_3}{\partial x} - \frac{\partial u_1}{\partial z},$$
$$Z = \frac{\partial u_1}{\partial y} - \frac{\partial u_2}{\partial x}.$$

These equations are so constituted that they satisfy equation (4) in section 77 by themselves. Equations (1), (2), and (3) in section 77 will now yield

$$(2)\quad \begin{aligned} \frac{\partial^2 u_1}{\partial y^2} + \frac{\partial^2 u_1}{\partial z^2} - \frac{\partial^2 u_2}{\partial y\,\partial x} - \frac{\partial^2 u_3}{\partial z\,\partial x} &= 4\pi i_1, \\ \frac{\partial^2 u_2}{\partial z^2} + \frac{\partial^2 u_2}{\partial x^2} - \frac{\partial^2 u_3}{\partial z\,\partial y} - \frac{\partial^2 u_1}{\partial x\,\partial y} &= 4\pi i_2, \\ \frac{\partial^2 u_3}{\partial x^2} + \frac{\partial^2 u_3}{\partial y^2} - \frac{\partial^2 u_1}{\partial x\,\partial z} - \frac{\partial^2 u_2}{\partial y\,\partial z} &= 4\pi i_3. \end{aligned}$$

Functions u_1, u_2, u_3 are not yet completely determined by means of these partial differential equations. However, assuming that one has found a solution for u_1, u_2, u_3, then one can designate $F(x, y, z)$ as any function of x, y, z which, with its derivatives, is finite and continuously variable. Then functions

$$u_1 + \frac{\partial F}{\partial x}, \quad u_2 + \frac{\partial F}{\partial y}, \quad u_3 + \frac{\partial F}{\partial z}$$

will also satisfy the partial differential equations in (2) and, by virtue of the equations in (1), these functions will also give the same result for X, Y, Z, as the solution for u_1, u_2, u_3 gives. The reverse also holds, for when one has found another solution U_1, U_2, U_3 in addition to solution u_1, u_2, u_3, then the differences, U_1-u_1, U_2-u_2, and U_3-u_3, can be assumed to be the partial differential derivatives of one and the same function $F(x, y, z)$ according to x, y, and z, respectively. What results from equations (2) for these differences is

$$\frac{\partial}{\partial y}\left\{\frac{\partial (U_1 - u_1)}{\partial y} - \frac{\partial (U_2 - u_2)}{\partial x}\right\} = \frac{\partial}{\partial z}\left\{\frac{\partial (U_3 - u_3)}{\partial x} - \frac{\partial (U_1 - u_1)}{\partial z}\right\},$$

$$\frac{\partial}{\partial z}\left\{\frac{\partial (U_2 - u_2)}{\partial z} - \frac{\partial (U_3 - u_3)}{\partial y}\right\} = \frac{\partial}{\partial x}\left\{\frac{\partial (U_1 - u_1)}{\partial y} - \frac{\partial (U_2 - u_2)}{\partial x}\right\},$$

$$\frac{\partial}{\partial x}\left\{\frac{\partial (U_3 - u_3)}{\partial x} - \frac{\partial (U_1 - u_1)}{\partial z}\right\} = \frac{\partial}{\partial y}\left\{\frac{\partial (U_2 - u_2)}{\partial z} - \frac{\partial (U_3 - u_3)}{\partial y}\right\}.$$

These partial differential equations are satisfied when one puts

$$\frac{\partial (U_1 - u_1)}{\partial y} = \frac{\partial (U_2 - u_2)}{\partial x},$$

$$\frac{\partial (U_2 - u_2)}{\partial z} = \frac{\partial (U_3 - u_3)}{\partial y},$$

$$\frac{\partial (U_3 - u_3)}{\partial x} = \frac{\partial (U_1 - u_1)}{\partial z}.$$

And what is expressed in this is that the differences U_1-u_1, U_2-u_2, and U_3-u_3 are the respective derivatives taken for x, y, z of one and the same function F(x, y, z).

So, in order to completely determine functions u_1, u_2, u_3, one can add another equation. We will select equation

$$\text{(3)} \quad \frac{\partial u_1}{\partial x} + \frac{\partial u_2}{\partial y} + \frac{\partial u_3}{\partial z} = 0$$

through which equations (2) will change into the following:

$$\frac{\partial^2 u_1}{\partial x^2} + \frac{\partial^2 u_1}{\partial y^2} + \frac{\partial^2 u_1}{\partial z^2} = 4\pi i_1,$$

$$\text{(4)} \quad \frac{\partial^2 u_2}{\partial x^2} + \frac{\partial^2 u_2}{\partial y^2} + \frac{\partial^2 u_2}{\partial z^2} = 4\pi i_2,$$

$$\frac{\partial^2 u_3}{\partial x^2} + \frac{\partial^2 u_3}{\partial y^2} + \frac{\partial^2 u_3}{\partial z^2} = 4\pi i_3.$$

These partial differential equations are satisfied by the solutions

$$u'_1 = -\int \frac{i_1\, dT}{r},$$

$$\text{(5)} \quad u'_2 = -\int \frac{i_2\, dT}{r},$$

$$u'_3 = -\int \frac{i_3\, dT}{r}.$$

Here, i_1, i_2, and i_3 signify the specific current intensities in point (x, y, z), dT is the spatial element contiguous to this point, and r is the distance of this same point from point (x′, y′, z′). u'_1, u'_2, u'_3 denote the values for u_1, u_2, u_3 in the latter point. One has to extend the integrations over all of the conductors that have current flowing through them.

Section 79
A Problem From The Theory Of Earth Magnetism

We will now deal with a problem that is of importance in the theory of earth magnetism.

There are magnetic masses present in the interior of a simply connected body, whose distribution is unknown. But, the components X, Y, Z, of the magnetic force exerted by these masses are known for every point in exterior space. These components are the partial derivatives of a potential function V, which is uniquely determined for every point in exterior space up to an additive constant. The value of the additive constant results from the condition that function V is zero at an infinite distance.

In exterior space function V, along with all of its derivatives, will be finite and continuous everywhere and it will satisfy the partial differential equation

$$(1) \quad \frac{\partial^2 V}{\partial x^2} + \frac{\partial^2 V}{\partial y^2} + \frac{\partial^2 V}{\partial z^2} = 0$$

every place in *exterior* space. Now function V can be continued into the interior of the given body in an infinite manifold of ways, that is, so that it will be finite and continuous in the interior and so that at every point on the surface it will assume the value given for that point. All such continuations will then furnish for an interior point in general another value for sum

$$\frac{\partial^2 V}{\partial x^2} + \frac{\partial^2 V}{\partial y^2} + \frac{\partial^2 V}{\partial z^2}.$$

This sum, divided by 4π, will give the magnetic density at the point in question. Thus, as one can see, there are infinitely many distributions of magnetic masses in the interior of the body constituted in such a way that they bring about the prescribed magnetic effects in all of the exterior space.

But now one can specifically look at the case where there are *no* magnetic masses and *no* galvanic current in all of the interior space as well as in the exterior space. Two questions arise from this, namely,

1) Can the prescribed magnetic effects in exterior space be produced in this case so that no galvanic current and only magnetic masses will be distributed in the body's surface?

2) Can those effects be brought about so that no magnetic masses and only galvanic current occur at the body's surface?

Each of these questions must be dealt with separately. It will be found that in each case one and only one specific distribution of magnetism or of current brings about the function in question.

First of all, the conditions have been set up which specify that there will be no magnetic masses and no galvanic current present in the body's *interior*. This will require that

$$(2) \quad \frac{\partial X}{\partial x} + \frac{\partial Y}{\partial y} + \frac{\partial Z}{\partial z} = 0$$

every place in the body's *interior* and that these three equations will likewise be satisfied at every place in the *interior:*

$$(3) \quad \frac{\partial Y}{\partial z} - \frac{\partial Z}{\partial y} = 0, \quad \frac{\partial Z}{\partial x} - \frac{\partial X}{\partial z} = 0, \quad \frac{\partial X}{\partial y} - \frac{\partial Y}{\partial x} = 0.$$

As a consequence of equations (2) and (3), X, Y, and Z in the body's interior are partial derivatives of a function V, namely,

$$(4) \quad X = \frac{\partial V}{\partial x}, \quad Y = \frac{\partial V}{\partial y}, \quad Z = \frac{\partial V}{\partial z},$$

and, in the body's *interior,* this function V satisfies partial differential equation (1).

We will denote S as the body's surface. A normal will be drawn at point (x, y, z) on this same surface towards the interior and the exterior. A distance p on this normal is positive towards the exterior and negative towards the interior. Respectively, p=+0, and p=-0 mean that this involves a point on the normal which, outside or inside the body, respectively, lies infinitely close to the surface. The value for function V and its first derivative in such a point can be denoted by attaching an index of +0 and -0, respectively. It is to be noted that V_{+0} and $(\partial V/\partial p)_{+0}$ are known for every point on surface S.

Section 80

Continuation: Simulated Distribution Of Magnetic Masses On The Surface Of A Magnet

What will be produced first are those given magnetic effects in exterior space by means of which only *magnetic masses are distributed on the body's surface, and by means of which no galvanic current enters the surface.*

This problem can be formulated in the following way.

Function V is given for every point in *exterior* space. There, it, along with all of its derivatives, is finite and continuous everywhere and satisfies partial differential equation (1) in section 79. Function V will be defined for the *interior* of the body in such a way that there it will satisfy the partial differential equation

$$(1)\quad \frac{\partial^2 V}{\partial x^2}+\frac{\partial^2 V}{\partial y^2}+\frac{\partial^2 V}{\partial z^2}=0,$$

that it, along with its derivatives, will be finite and continuous in the interior, and that

$$(2)\quad V_{-0}=V_{+0}$$

every place on the surface.

Equation (1) states that there are no magnetic masses and no current in the interior, and equation (2) states that there is no current present in the surface.

This problem has been solved in section 21 and it has been proven in section 34 that it always has one and only one solution. If one has found it, then, according to equation (9) in section 65, the density of the magnetic masses at a point (x, y, z) results from the equation

$$(3)\quad \left(\frac{\partial V}{\partial p}\right)_{+0}-\left(\frac{\partial V}{\partial p}\right)_{-0}=4\,\pi\,\rho.$$

Section 81

Continuation: Simulated Galvanic Current On The Surface Of The Magnet

We now proceed to the second problem. Those given magnetic effects in exterior space will be produced so that galvanic current will only *occur on the body's surface and no magnetic masses will be present anywhere.*

This problem is formulated in the following way.

Function V is given for every point in exterior space in the same way as it was in the previous problem. Its continuation will be defined for the interior of the body in such a way that there it will satisfy the partial differential equation

$$(1)\quad \frac{\partial^2 V}{\partial x^2}+\frac{\partial^2 V}{\partial y^2}+\frac{\partial^2 V}{\partial z^2}=0,$$

that it, along with its derivatives, will be finite and continuous in the interior, and that

$$(2)\quad \left(\frac{\partial V}{\partial p}\right)_{-0} = \left(\frac{\partial V}{\partial p}\right)_{+0} = N$$

every place on the surface.

Equation (1) states that there are no magnetic masses and no current in the interior, and equation (2) states that no magnetic masses are present on the surface.

It further follows from equation (1) that

$$(3)\quad \int N\, d\sigma = 0,$$

when the integral is extended across the body's surface. In order to prove this, we will set up a normal towards the interior at point (x, y, z) on the surface and will denote a distance on the normal from that point out as n, so that n= -p for negative values of p. By applying the lemma we developed in equation (4) of section 19, one obtains

$$\int\int\int\left(\frac{\partial X}{\partial x} + \frac{\partial Y}{\partial y} + \frac{\partial Z}{\partial z}\right) dx\, dy\, dz = -\int\left(X\frac{\partial x}{\partial n} + Y\frac{\partial y}{\partial n} + Z\frac{\partial z}{\partial n}\right) d\sigma.$$

The integral on the left here is to be extended over the entire body, and the integral on the right is to be extended over its surface. And if one considers equations (4) in section 79, then the last equation can be written as

$$\int\int\int\left(\frac{\partial^2 V}{\partial x^2} + \frac{\partial^2 V}{\partial y^2} + \frac{\partial^2 V}{\partial z^2}\right) dx\, dy\, dz = -\int\frac{\partial V}{\partial n}\, d\sigma.$$

But, according to equation (1), the left-hand side is equal to zero. Furthermore:

$$(4)\quad -\frac{\partial V}{\partial n} = \left(\frac{\partial V}{\partial p}\right)_{-0} = N.$$

If one substitutes this, then one obtains the equation to be proven in (3).

The point is now to prove that there is always one and *only* one function V, which satifies the established conditions.

With this end in mind, we will denote by u a single-valued function of x, y, z about which nothing else will be stipulated other than that it and its first derivatives should be finite and continuous everywhere in the interior of the body. There are infinitely many such functions. Consequently, integral

(5) $$\Omega(u) = \int\int\int \left\{ \left(\frac{\partial u}{\partial x}\right)^2 + \left(\frac{\partial u}{\partial y}\right)^2 + \left(\frac{\partial u}{\partial z}\right)^2 \right\} dx\, dy\, dz,$$

which is extended over the body, can assume an infinite number of different values. But no matter which function u may be chosen, the value for $\Omega(u)$ will always turn out positive and finite. The first results from the form of the integral, and the other directly follows from the assumption. We will only want to take those functions u into consideration of which no two have a constant ratio to each other. This limitation is introduced through the secondary condition of having the surface integral

(6) $$\int u\, N\, d\sigma = A.$$

We will understand A to be a constant different from zero, whose magnitude can remain provisionally undetermined.

If we designate v as one of the infinitely many functions u, then all the other functions can be written in the form $u=v+hs$, when h signifies an appropriately selected constant, s is a function that has the same condition attached to it in the interior of the body as do the functions u, and there is the secondary condition originating from equation (6) so that surface integral

(7) $$\int s\, N\, d\sigma = 0.$$

Because the values of $\Omega(u)$ are finite and positive, there exists among the integrals (5) for which the secondary condition (6) is satisfied at least *one* minimum. We will denote by v the function which attains this minimum. Then, all the other functions u can be written in the form $u=v+hs$, and if one assumes h to be infinitely small, then the condition for the minimum will be

(8) $$\Omega(v) \leqq \Omega(v + h\, s).$$

Integral $\Omega(v+hs)$ can be developed in the same way as in section 34. We obtain

(9) $$\Omega(v + h\, s) = \Omega(v) + 2h \int\int\int \left(\frac{\partial v}{\partial x}\frac{\partial s}{\partial x} + \frac{\partial v}{\partial y}\frac{\partial s}{\partial y} + \frac{\partial v}{\partial z}\frac{\partial s}{\partial z} \right) \times dx\, dy\, dz + h^2\, \Omega(s).$$

We will want to transform the second component on the right-hand side of this equation according to section 20. It will turn out to be

$$\int\int\int\left(\frac{\partial v}{\partial x}\frac{\partial s}{\partial x}+\frac{\partial v}{\partial y}\frac{\partial s}{\partial y}+\frac{\partial v}{\partial z}\frac{\partial s}{\partial z}\right)dx\,dy\,dz$$

(10)
$$=-\int s\frac{\partial v}{\partial n}\,d\sigma-\int\int\int s\left(\frac{\partial^2 v}{\partial x^2}+\frac{\partial^2 v}{\partial y^2}+\frac{\partial^2 v}{\partial z^2}\right)dx\,dy\,dz.$$

However, only those functions s will be considered that satisfy the secondary condition (7). In order to take these into account, we will multiply both of its sides by a temporarily still-undetermined constant k, then by 2h, and we will combine the result with equation (10) through addition. What will result from this is that, in case of the validity of equation (7), equation (9) will change into the following:

$$\Omega\,(v+h\,s)\;=\;\Omega\,(v)-2\,h\int s\left(\frac{\partial v}{\partial n}-k\,N\right)d\sigma$$

(11)
$$-\,2\,h\int\int\int s\left(\frac{\partial^2 v}{\partial x^2}+\frac{\partial^2 v}{\partial y^2}+\frac{\partial^2 v}{\partial z^2}\right)dx\,dy\,dz$$

$$+\,h^2\,\Omega\,(s).$$

This holds for every value of the constant h. So, in order for condition (8) to be satisfied for an infinitely small h, the aggregate of what is multiplied by 2h on the right-hand side of equation (11) must be made equal to zero. To have this, it is necessary and sufficient that

(12) $$\frac{\partial^2 v}{\partial x^2}+\frac{\partial^2 v}{\partial y^2}+\frac{\partial^2 v}{\partial z^2}=0$$

every place in the interior of the body and that

(13) $$\frac{\partial v}{\partial n}=k\,N$$

for every point on its surface.

If we then put

(14) $$V=-\frac{1}{k}\,v,$$

then function V satisfies all the established conditions.

Constant k is dependent on the value of A, but, in any case, it is different from zero. For, if one were to assume that $k=0$ because of equation (13), $\partial v/\partial n=0$ would have to hold at every point on the surface. So considering this and equation (12), one would get

$$-\int v\,\frac{\partial v}{\partial n}\,d\sigma-\int\int\int v\left\{\frac{\partial^2 v}{\partial x^2}+\frac{\partial^2 v}{\partial y^2}+\frac{\partial^2 v}{\partial z^2}\right\}dx\,dy\,dz\;=\;0.$$

But the left-hand side of this equation results from integral

$$\int\int\int\left\{\left(\frac{\partial v}{\partial x}\right)^2+\left(\frac{\partial v}{\partial y}\right)^2+\left(\frac{\partial v}{\partial z}\right)^2\right\} dx\, dy\, dz$$

by means of the transformation in section 20, and this integral would also have to have a value of zero, which is not otherwise possible than when one puts v=const. What would immediately result from this is $\int v N\, d\sigma = c\int N\, d\sigma = 0$ according to equation (3), and this would contradict secondary condition (6). Therefore, k must be different from zero.

It still remains to be proven that there is no other function besides v under which integral (5) attains a minimum given condition (6). Assuming that u=v+s would be such a function, then it would satisfy the condition

(15) $\Omega\,(v+s) \leqq \Omega\,(v+h\,s),$

in which constant h here is assumed to be infinitely close to one. But, according to equations (11), (12), and (13), $\Omega(v+hs)=\Omega(v)+h^2\Omega(s)$, $\Omega(v+s)=\Omega(v)+\Omega(s)$. And consequently, condition (15) will now read

(16) $\Omega\,(s) \leqq h^2\,\Omega\,(s).$

But one can take constant h^2, which is supposed to lie infinitely close to one here, to be not just simply greater than one, for it can also be taken to be smaller than one, and therefore condition (16) can only be satisfied by putting $\Omega(s)=0$, i.e., s= const. So, if one disregards an arbitrary additive constant, v is, therefore, the only function under which integral (5) attains a minimum in keeping with the secondary condition in (6).

Finally, one can still investigate the connection between k and A. It has already been proven that

$$\int\int\int\left\{\left(\frac{\partial v}{\partial x}\right)^2+\left(\frac{\partial v}{\partial y}\right)^2+\left(\frac{\partial v}{\partial z}\right)^2\right\} dx\, dy\, dz$$
$$= -\int\int\int v\left(\frac{\partial^2 v}{\partial x^2}+\frac{\partial^2 v}{\partial y^2}+\frac{\partial^2 v}{\partial z^2}\right) dx\, dy\, dz - \int v\,\frac{\partial v}{\partial n}\, d\sigma.$$

The triple integral on the left is the minimum $\Omega(v)$. The triple integral on the right will drop out due to equation (12). If one now also brings equation (13) into play, the last equation will change into $\Omega(v)=-k\int vN d\sigma$, i.e., into $\Omega(v)=-kA$. Because we can still dispose of A, we can put $A=\Omega(v)$ and thus obtain from equation (14):

(17) $V = v.$

So, disregarding an additive constant, there is only one function V that satisfies the conditions established at the beginning of this chapter.

The theorem, then, is only valid when the given body is simply connected.

Section 82
Continuation: Streamlines

Although one has defined function V for the interior of the body in the treatment of the problem in section 81, what still remains is to investigate the current in surface S. For this purpose, we will draw an arbitrary curve that does not bisect itself from point 1 to point 2 in the surface. [Figure 43] We will consider the positive side to be that which lies on a person's left hand when he runs around the curve from 1 to 2 on the outside of surface S. We will furthermore set up a bounded surface T which bisects surface S along curve 1 2. Curve 1 2 will break down surface T into two separate segments, of which one will be outside and the other inside the given body. We will describe a positive circuit around T's boundary as that which leads from 1 to 2 outside the body, and from 2 to 1 inside the body.

In order to find J, the quantity of electricity of which more streams over through curve 1 2 from its negative to its positive side in the time unit than goes in the reverse direction, we will have to extend integral $1/4\pi \int dV$ in a positive direction through T's boundary according to section 75. The path of integration of 1 2 *outside* of the body will be

$$\int dV = (V_2)_{p=+0} - (V_1)_{p=+0},$$

and the path of integration for 2 1 *inside* will be

Figure 43

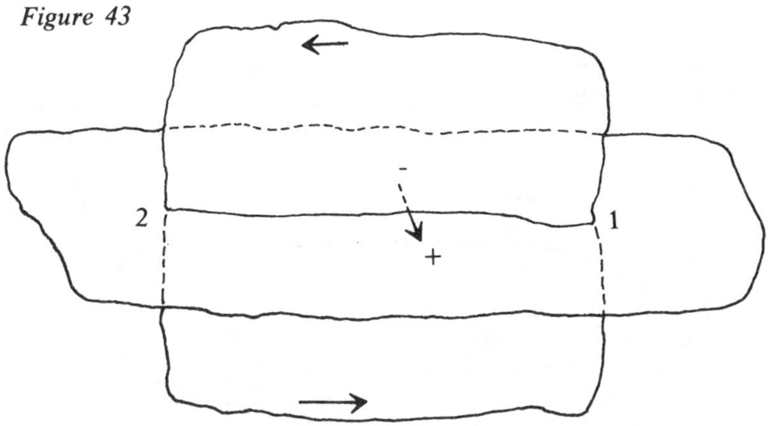

$$\int dV = (V_1)_{p=-0} - (V_2)_{p=-0}.$$

So, if we have

(1) $V_{+0} - V_{-0} = 4\pi W,$

then we will obtain

(2) $J = W_2 - W_1.$

One can now draw a system of self-enclosing lines in surface S [Figure 44] so that, in one and the same line, W has a constant value that changes when one crosses over from one line to another. Just as much electricity will flow from one side to the other through such a line during the time unit as will flow in reverse. This turns out to be the same thing as if one had said: "No electricity at all flows through such a line." These lines, then, are the streamlines. For any two such lines: $W = W_1$, $W = W_2$, giving the difference $W_2 - W_1$, which is the *current intensity* of the electricity moving between them.

It is to be noted that function W, just like function V in section 81, is subjected to an arbitrary additive constant. However, this constant does not exert any influence at all on the difference $W_2 - W_1$ because it is the same in its minuend and subtrahend. So, there is only *one* system of streamlines and there is only one determinate current intensity between any two lines in this system.

If we take an infinitely small line element ds on surface S, then integral

$$W_{s+ds} - W_s = \frac{\partial W}{\partial s} ds$$

is the quantity of electricity which crosses over perpendicular to the line element from its negative side to its positive in the unit of time, reduced by the quantity of electricity which goes through in the opposite direction in

Figure 44

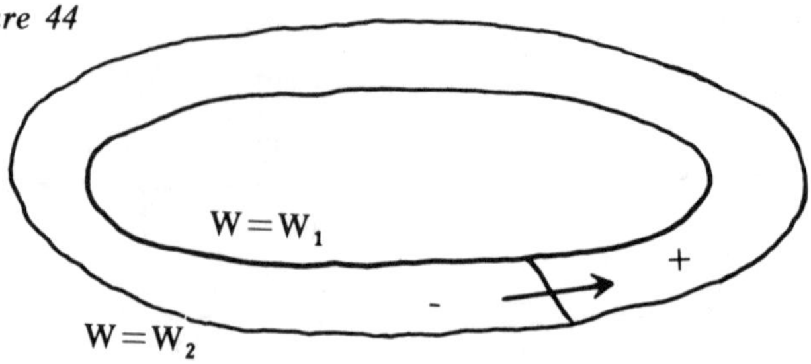

the same unit of time. We will call this magnitude, divided by ds, the *specific current intensity* in the direction perpendicular to the line element. If we denote this as i, then

$$(3) \quad i = \frac{\partial W}{\partial s}.$$

The earth is a simply connected body. Thus, the investigations in sections 79-82 can be applied to it. In order to trace back the magnetic effects that are observed in exterior space to their origin, one can assume that there are infinitely many various distributions of magnetic masses in the interior of the earth. But these can also be replaced by a single distribution of magnetic fluids on the surface or by a single arrangement of galvanic current on the surface.

This distribution of magnetic fluids on the earth's surface is only an ideal one. But, on the other hand, in actuality there can be galvanic currents present in this surface. If one wanted solely to explain exterior magnetic effects as coming from magnetic masses in the interior of the earth, then these same masses would have to be present in enormous quantities. On the other hand, very weak galvanic currents in this surface would be sufficient to produce those exterior magnetic effects. This, however, would presuppose that we can prove that there exists a permanent electromotive force for these currents.

Section 83

Multiply Connected Bodies

We will now want to consider a multiply connected body more precisely. A body is called simply connected if it does not allow a sectional plane to go through it in any way that would not break it down into completely separate pieces. If a body allows itself to be changed into a simply connected one by means of q number of sectional planes, then we will call it a (q+1)-fold connected body. Exterior space will also be multiply connected for such a body; it will even be (q+1)-fold in the case where the body itself lies totally bounded in a finite domain.

In order to prove this theorem, we first of all have to set up the schema of a multiply connected body. The system of wires in section 62 presents itself as such a schema if we provide that the cross-section of all the wires will be finite in two dimensions (so that we will not have to consider a special case). We will number all of the m individual wire branches and all of the n nodules. The proof that we are conducting will be based on enumerations that can be worked out for the sake of

Figure 45

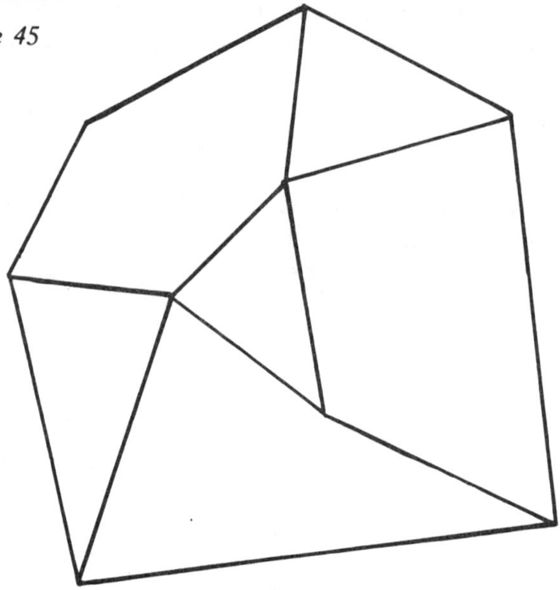

simplicity on a plane figure. We will assume there to be just as many points in the plane at a finite distance to each other as there are nodules in the wire system. Each point in the plane and its similarly numbered nodule in the wire system will correspond to each other. The points on the plane will be connected only by means of straight lines in such a way that each such line will correspond to a specific, single wire branch with the reverse relationship valid too, and each beginning and end point, respectively, of any line will correspond to the nodules between which lies the wire branch corresponding to the line. Because nothing is presumed about the mutual position of the nodal points in the plane, one can always arrange them in such a way that none of the connecting lines are bisected by another one between their beginning and end points, and so that no two lines that meet in a nodal point will form an angle of precisely 180 degrees. [Figure 45]

The plane figure consists of a polygon, all of whose corner points are nodal points and which contains the remaining nodal points in its interior. A portion of the drawn lines forms the polygon's boundary, while the remainder lie in its interior and break it down into a number of individual figures which are still to be determined more precisely.

If one then cuts every line in Figure 45 at a place lying between the nodal points and prescribes that a continuous circuit of the linear system cannot cross any cut, then this same system will break down into an n-fold

Figure 46

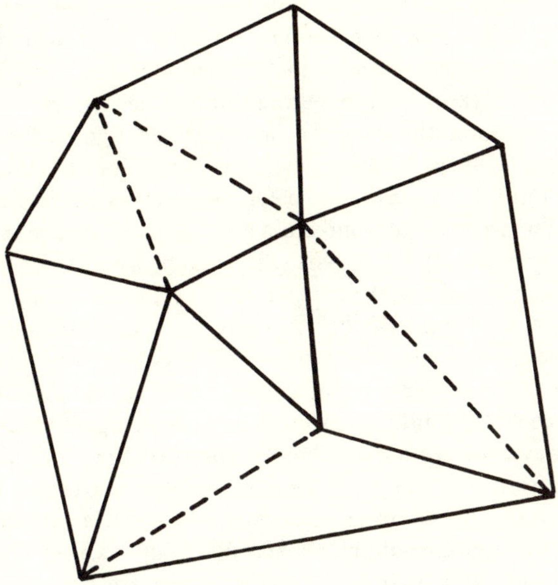

connected system, in which each section will contain a nodal point. In actuality, one can continuously move along from every nodal point out to all the lines spreading out from it, but only up to the cut. If one now allows one cut to drop out, then both simply connected systems which touch each other there will unite into a single simply connected system. Or, in other words: by eliminating a cut, the number pertaining to the simply connected linear system is reduced by one. So, if one now wants to maintain a *single* simply connected system, one must eliminate n - 1 cuts. The given system of lines will then be changed into a simply connected system by means of m - n + 1 cuts in the interior of the lines.

Now, in Figure 45, there are a-number of exterior nodal points present, and consequently there are also a-number of exterior boundary lines. In order to determine the number of the individual figures into which the a-gon will be broken down by interior lines, we will draw diagonals out from a point in every figure that has more than three sides. (They are the dotted lines in Figure 46.) As a result, the inner figures will break down into genuine triangles whose number can be easily determined from their angular sum. Namely, the sum of the angles at the exterior nodal points is (a - 2) 2R and at the inner nodal points (n - a) 4R. Consequently, we have 2(n - 1) - a *triangles* and 6(n - 1) - 4a *interior* triangle sides. Every interior line in Figure 46 has been counted twice in this. So, the number of these interior lines is 3(n-1)-2a, while, in the original Figure 45, there are only m-a interior lines present. Consequently, in Figure 46, we have 3(n-1)-a-m dotted lines, which will have to be removed if one wants to return to the original figure. Two adjoining figures will be united into one for every dotted line removed. Therefore, the number of the single figures into which the a-gon in Figure 45 is broken down by means of the interior line is $2(n-1)-a-[3(n-1)-a-m]=m-n+1$. Now, every cut in the interior of the lines in Figure 45 corresponds to a cross-section surface in the interior of the given wire system. Thus, this same system is converted into a simply connected body by means of $m-n+1$ cross-section surfaces. Every single simple figure in Figure 45 corresponds to a cross-section surface in exterior space. Therefore, the number of these cross-section surfaces is also $m-n+1$. If all these cross-section surfaces are present, then the exterior space is still totally connected. For one can get from one point on one side of any cross-section to all other points on that one side, as well as to the other side of every cross-section, without leaving exterior space. But, if one wanted to set up another new cross-section surface in exterior space, then this space would be broken down into two separate pieces by that. Thus, the given body and exterior space are both $(m-n+2)$-fold connected.

One can also see at the same time, from the development of the proof, that the given body can be broken down in a manifold of ways into being simply connected. But, the number of the cross-sections will be the same in all the dissections.

Having finished this interpolation, we will now return to the investigation in section 81.

Section 84

The Problem Of Section 81 For A Multiply Connected Body

The problem of section 81 will now be dealt with under the provision that the given body is $(q+1)$-fold connected.

First of all, we will break down the exterior space by means of q cross-section surfaces $S_1, S_2, \ldots S_q$ into a simply connected space and stipulate that all displacements that were undertaken in exterior space with a point (x, y, z) will lie completely inside this simply connected space, i.e., that no displacement will be allowed to cut through the surface of the given body or through any one of the cross-section surfaces $S_1, S_2, \ldots S_q$. According to (4) in Section 79, $X\,dx+Y\,dy+Z\,dz=dV$ is a complete differential everywhere in exterior space. If one then extends integral

$$(1)\quad V=\int(X\,dx+Y\,dy+Z\,dz)$$

from an infinite distance towards point (x, y, z) lying in exterior space, then the value of this integral will be independent of the path of integration as long as this course, in terms of its entire extension, lies in the simply connected exterior space. Consequently, function V inside of the stated space will be a single-valued, everywhere finite function of location, whose values will continuously change with every permissible continuous displacement of point (x, y, z).

Function V will have values with a finite difference for two points which lie infinitely close to each other on opposite sides of any one of the cross-sections. One can find this difference when one extends integral $\int(X\,dx+Y\,dy+Z\,dz)$ from the one point to the other along a path of integration which lies completely inside the simply connected exterior space. For one and the same cross-section the difference $V_{+0}-V_{-0}$ is constant no matter at what place of this cross-section the two infinitely close points are chosen. For if one draws two lines lying infinitely close to each other along a cross-section, with one line lying on the positive side and the other on the negative side of the cross-section, then in any point along one line components X, Y, Z, respectively, will have the same values as in the point lying infinitely close on the other line.

The q constant values, which the difference $V_{+0}-V_{-0}$ possesses on both sides of cross-section $S_1, S_2, \ldots S_q$, are known due to the nature of the problem. We will denote them as $C_1, C_2, \ldots C_q$.

If the function is now continued into the interior of the given body according to conditions (1) and (2) in section 81, then once again one can follow precisely the same path that was followed there. But the function which resulted there (it is denoted here as V′) is *no longer* the *single* solution to the problem. Namely, one can add another function V′′ which will contain q arbitrary constants $c_1, c_2, \ldots c_q$ as linearly occurring factors. What we will establish by q cross-sections $Q_1, Q_2, \ldots Q_q$ is simple connectedness in the interior of the given body and we will then go on to determine function v_μ according to the following conditions:

$$(2)\quad \frac{\partial^2 v_\mu}{\partial x^2}+\frac{\partial^2 v_\mu}{\partial y^2}+\frac{\partial^2 v_\mu}{\partial z^2}=0$$

in the interior.

$$(3)\quad \frac{\partial v_\mu}{\partial n}=0$$

for every point in the surface of the q-fold connected body. Here, n means the normal to this point drawn towards the interior.

(4) $(v_\mu)_{+0} - (v_\mu)_{-0} = 1$

for every two points, which lie infinitely close to each other on opposite sides of cross-section Q_μ. Finally,

(5) $\left(\frac{\partial v_\mu}{\partial P}\right)_{+0} = \left(\frac{\partial v_\mu}{\partial P}\right)_{-0}$

for this same cross-section, when we denote P as a distance drawn from a point on cross-section Q_μ out along the normal and which is positive on one side and negative on the other.

In the remainder of the problem, function v_μ along with its derivatives, will be finite and continuously variable everywhere inside of the entire q-fold connected space into which the given (q+1)-fold connected body can be changed by means of cross-section Q_μ.

This problem is solved in section 60. One only needs to take the magnitude k which occurs there as constant on both sides of surface Q_μ and throughout the entire body. It has furthermore been proven that the problem only allows *one* solution. So, if one now successively takes $\mu = 1$, 2, ... q, then one gets q different functions $v_1, v_2, v_3, \ldots v_q$, through which equations (2), (3), (4), and (5) are satisfied. The discontinuity prescribed in (4) occurs for every function only in one of the cross-sections $Q_1, Q_2, Q_3, \ldots Q_q$ and in a particular cross-section.

We then put

(6) $V'' = c_1 v_1 + c_2 v_2 + \ldots + c_q v_q$

in the given body's interior. Because function V, which is given in exterior space, has already totally made its influence felt (if one is allowed to express himself this way) in establishing V′ according to section 81, then

(7) $V'' = 0$

must necessarily be chosen for exterior space. Therefore, function

(8) $V = V' + V''$

satisfies the following conditions. In all exterior space it agrees with function V, which is given for that space. It satisfies partial differential equation (2) in exterior space, as well as in the interior of the body.

(9) $\left(\frac{\partial V}{\partial p}\right)_{-0} = \left(\frac{\partial V}{\partial p}\right)_{+0}$

for every point on the surface of the (q+1)-fold connected body. In transit through cross-sections $Q_1, Q_2, \dots Q_q$ from the negative to the positive side, function V changes discontinuously around the constant magnitude $c_1, c_2, \dots c_q$. Furthermore, along with its first derivatives, it is finite and continuously variable in the interior of a simply connected body, which is produced by means of cross-sections $Q_1, Q_2, \dots Q_q$. Its derivative in the direction of the increasing normal will have the same value for two points lying infinitely close to one another on different sides of cross-section Q.

Because coefficients $c_1, c_2, \dots c_q$ are completely arbitrary, there are infinitely many distributions of current in the surface of a multiply connected body, each of which is producing the prescribed magnetic effects in exterior space.

If one has assumed a definite system of constants $c_1, c_2, \dots c_q$, then the streamlines and the current intensities would result from an application of the procedure developed in section 82. One would then have to differentiate between two kinds of current. For, if one were to put constants $c_1, c_2, c_3, \dots c_q$ equal to zero, from which it would follow that $V''=0$, then one would get a single arrangement of current from which alone the exterior magnetic effects would originate. But, on the other hand, if one has $V'=0$, then, for every definite system of constants $c_1, c_2, \dots c_q$ one would get a distribution of current which would not exert any magnetic effects at all in exterior space. This would follow directly from equation (7). In section 85, we will want to consider the simplest case, namely, where $q=1$, even more closely.

Section 85

Continuation: The Ring

Let the given body be a ring and so twofold connected. [Figure 47] We will break it down by a cross-section Q and similarly its exterior space by a cross-section S into a simply connected space. As always, the boundary lines for cross-sections Q and S will lie on the ring's surface. There will be two self-enclosing lines, which will bisect each other at a point. Q's boundary line is constructed in such a way that every surface that it serves as a total boundary cuts the ring's axis at a point. On the other hand, surface S can be continued across its boundary into the ring's interior in such a way that the ring's axis lies totally in this continuation. One can now draw two systems of self-enclosing lines on the ring's surface, so that the lines for one and the same system lie completely separate from each other, while, on the other hand, every line in the first system bisects the lines of the second system at a point each. The systems will be

Figure 47

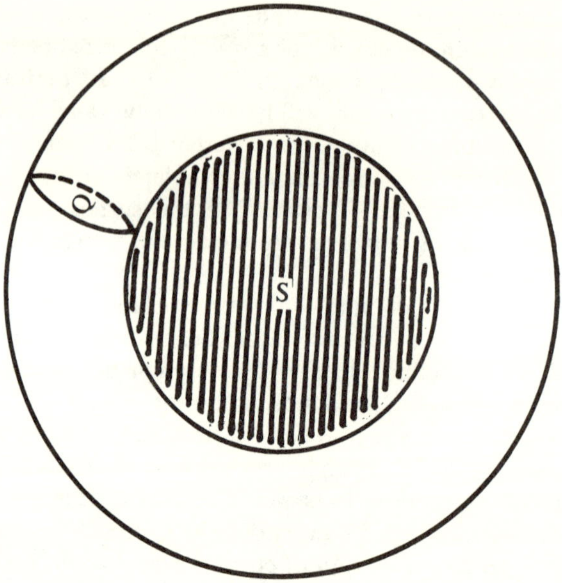

constructed in such a way that every two adjoining lines in the same system will lie infinitely close to each other, and S's boundary belongs to the first system, while Q's boundary belongs to the second system.

We will take two points which lie infinitely close to each other on opposite sides of surface S and we will connect them by means of a line that totally runs inside of the simply connected exterior space. One can turn this line into the boundary of a surface which will bisect the ring's surface in any line of the second system. We will position ourselves on that side of the surface in which a positive circuit through the boundary will lead from the negative to the positive side of S. In order to find J′, the quantity of electricity, of which more flows from below to above in the time unit through the two-dimensional surface set-up than flows from above to below, we will have to have integral $1/4\pi \int dV$ extend through the boundary line according to section 75. Specifically, it will extend from S's negative to its positive side. This integral's value is

$$\frac{1}{4\pi}(V_{+0} - V_{-0}).$$

But now let $V = V' + V''$ in the exterior space as in the ring's interior and notice that $V' = V$, $V'' = 0$ in all the exterior space. One therefore obtains the equation

$$(1)\quad J' = \frac{1}{4\pi}(V'_{+0} - V'_{-0}) = \frac{C}{4\pi}.$$

This equation would remain unchanged if one were to have $V''=0$ *everywhere.* Thus, if a current to which function V'' belongs were to at one place go through a line of the second system, then it would return again to the same place or to a different place on the original side. Therefore, the lines of the second system on the ring's surface (and cross-section Q's boundary with them) can be arranged in such a way that they will become the streamlines of the second kind of current. Their equations are contained in the general form

$$(2)\quad 4\pi W'' = -V''_{-0} = const.,$$

with V''_{-0} meaning the value for function V'' in the ring's interior infinitely close to its surface.

In the same manner, one finds that the lines of the first system, properly arranged, are the streamlines for the first kind of current. They will be determined by means of equations of the form

$$(3)\quad 4\pi W' = V'_{+0} - V'_{-0} = const.,$$

in which V'_{+0} and V'_{-0} are the value for V' in two points that lie infinitely close to each other on the exterior and interior side of the ring's surface.

The magnetic effects in exterior space will then originate solely in the currents which flow in the paths in equation (3). The currents, whose streamlines belong to equation (2), will not exert any magnetic effects in exterior space.

Section 86

The Magnetic Potential

We can determine the interaction of two permanent magnets by establishing their potential on each other. Let there be a magnetic mass $d\mu$ in any spatial element of the first magnet and magnetic mass $d\mu'$ in a spatial element of the second magnet. These masses $d\mu$ and $d\mu'$ will be independent of time t. Then,

$$(1)\quad V = -\int \frac{d\mu}{r}$$

is the potential function of the first magnet acting on the positive magnetic unit which is concentrated in the previously mentioned point (x, y, z). Furthermore,

(2) $V' = -\int \frac{d\mu'}{r}$

is the potential function of the second magnet acting on the positive magnetic unit which is concentrated in the previously mentioned point (x, y, z). Here, r will denote the distance of a point in the spatial element filled with magnetic mass $d\mu$ or $d\mu'$, respectively, from point (x, y, z). The integration in (1) extends across the entire first magnet, and in (2) across the entire second magnet.

The potential P of both magnets on each other is expressed by equation

(3) $P = -\int\int \frac{d\mu \cdot d\mu'}{r}.$

Here, r signifies the distance between two points, one of which belongs to the spatial element filled with $d\mu$, and the other to the spatial element filled with $d\mu'$. The integration in (3) is to be extended over both magnets. By comparing formulas (1), (2), and (3), one can recognize that P can be expressed in two ways:

(4) $P = \int V' \, d\mu,$

(5) $P = \int V \, d\mu'.$

Equation (4) is to be understood in such a way that point (x, y, z) occurring in (2) is transferred into the first magnet's spatial element that is filled with $d\mu$. On the other hand, in (5), one has to transfer point (x, y, z) occurring in (1) into the second magnet's spatial element that is filled with $d\mu'$. In (4), the integral extends over the first magnet and in (5), it extends over the second magnet.

Section 87

The Electromagnetic Elementary Work

We will next want to go into the electromagnetic interaction between a constant linear galvanic current and a magnet. We will only need to construct a surface S whose boundary is the current's path and this surface, as well as one lying infinitely close to it, is subjected to a magnetic mass according to section 72. As a result of this, we will obtain a magnet which exerts the same magnetic effects toward the outside as does the given current.

At a point on surface S, we will set up a normal on one side, and denote p as a distance counted off on it from that point. One then has to put

$$d\mu = \frac{J}{\delta}\, d\sigma \quad \text{for } p=0,$$

$$d\mu = -\frac{J}{\delta}\, d\sigma \quad \text{for } p=\delta.$$

J here signifies the intensity of the linear current, δ is an infinitely small length, and $d\sigma$ is an element of surface S.

So, with V′ being the potential function of the given magnet, we then have

$$(1)\quad \begin{aligned} P &= \int V'\, d\mu = \int V'_0 \cdot \frac{J}{\delta}\, d\sigma - \int V'_\delta \frac{J}{\delta}\, d\sigma \\ &= \int \frac{V'_0 - V'_\delta}{\delta} \cdot J\, d\sigma = -\int \frac{\partial V'}{\partial p} J\, d\sigma. \end{aligned}$$

Instead of J, we could further introduce the potential function V of the magnetic force exerted by the galvanic current. According to section 71, it is the magnetic measurement $4\pi J = V_{+0} - V_{-0}$. Consequently, we get

$$(2)\quad P = -\frac{1}{4\pi} \int (V_{+0} - V_{-0}) \frac{\partial V'}{\partial p}\, d\sigma.$$

The value for P will change with an infinitely small displacement of the magnet. The change gives the work that the magnetic forces exerted by the current have to produce in order to bring about that displacement. Conversely, the effect of the magnet on the current is found according to the theory of the equality of action and reaction.

Section 88

The Electromagnetic Elementary Work. Two Constant Linear Currents

In section 86, we considered the interaction between two magnets. In section 87, a constant galvanic current was used in place of the first magnet and one can also use a constant current in place of the other magnet. The question here is the interaction between two constant currents. Insofar as the work performed by this is used to move the current *with* the current conductors, we will call this interaction the *electrodynamic* interaction.

A function P will now be set up whose infinitely small change indicates the electrodynamic elementary work that is performed by an infinitely small displacement of both currents.

We can therefore proceed from equation (2) in section 87, but we will now have to consider V′ as the potential function of the magnetic force which is exerted by a linear galvanic current. In point (x, y, z), the components of this force will be

$$(1)\quad \frac{\partial V'}{\partial x} = X', \quad \frac{\partial V'}{\partial y} = Y', \quad \frac{\partial V'}{\partial z} = Z',$$

and it is to be noted that X′, Y′, Z′ are finite and continuously variable everywhere outside of the linear currents from which they originate. What results is

$$\frac{\partial V'}{\partial p} = X' \frac{\partial x}{\partial p} + Y' \frac{\partial y}{\partial p} + Z' \frac{\partial z}{\partial p},$$

and consequently, one can now write equation (2) of section 87 thus:

$$(2)\quad \begin{aligned} P = &-\frac{1}{4\pi}\int V_{+0}\, d\sigma \left\{ X' \frac{\partial x}{\partial p} + Y' \frac{\partial y}{\partial p} + Z' \frac{\partial z}{\partial p} \right\} \\ &+\frac{1}{4\pi}\int V_{-0}\, d\sigma \left\{ X' \frac{\partial x}{\partial p} + Y' \frac{\partial y}{\partial p} + Z' \frac{\partial z}{\partial p} \right\}. \end{aligned}$$

But, if we consider both sides of surface S as a segment of the boundary of infinite space (the other boundary is an infinitely distant spherical surface), then one can draw normal n towards the interior of this space from the positive side as well as from the negative side of S. On the side of positive p, we have n=p and, on the other side, n=-p. Therefore, equation (4) will now be

$$(3)\quad P = -\frac{1}{4\pi}\int V\, d\sigma \left\{ X' \frac{\partial x}{\partial n} + Y' \frac{\partial y}{\partial n} + Z' \frac{\partial z}{\partial n} \right\},$$

when the integral is extended over *both* sides of surface S.

This integral can be replaced by a volume integral. Namely, if we denote T as an infinite space that has an infinitely distant spherical surface and both sides of surface S as its boundary, then what occurs according to (4) in section 19 is that integral

$$\int dT \left(\frac{\partial (V X')}{\partial x} + \frac{\partial (V Y')}{\partial y} + \frac{\partial (V Z')}{\partial z} \right),$$

which is extended over the infinite space, is equal to the surface integral

$$-\int V\, d\sigma \left(X' \frac{\partial x}{\partial n} + Y' \frac{\partial y}{\partial n} + Z' \frac{\partial z}{\partial n} \right),$$

when this is extended over both sides of surface S and over the infinitely distant spherical surface. But now, at an infinite distance, X′, Y′, Z′, as well as V, are equal to zero. So, the integral that is extended over the spherical surface can drop away and we obtain

$$(4)\quad P = \frac{1}{4\pi}\int dT\left\{\frac{\partial(VX')}{\partial x} + \frac{\partial(VY')}{\partial y} + \frac{\partial(VZ')}{\partial z}\right\}.$$

The integration in equation (4) is to be extended over all of infinite space.

We can transform further. Namely, what results by carrying out the differentiation is

$$\frac{\partial(VX')}{\partial x} + \frac{\partial(VY')}{\partial y} + \frac{\partial(VZ')}{\partial z} = X'\frac{\partial V}{\partial x} + Y'\frac{\partial V}{\partial y} + Z'\frac{\partial V}{\partial z} + V\left(\frac{\partial X'}{\partial x} + \frac{\partial Y'}{\partial y} + \frac{\partial Z'}{\partial z}\right).$$

But, because X′, Y′, Z′ originate in a linear galvanic current,

$$\frac{\partial X'}{\partial x} + \frac{\partial Y'}{\partial y} + \frac{\partial Z'}{\partial z} = 0$$

in all the infinite space outside of the current's conductor. Furthermore, V is the potential function of the magnetic force that is exerted by the first linear galvanic current; consequently,

$$\frac{\partial V}{\partial x} = X,\quad \frac{\partial V}{\partial y} = Y,\quad \frac{\partial V}{\partial z} = Z.$$

Therefore, instead of equation (4), one can also put

$$(5)\quad P = \frac{1}{4\pi}\int dT\,(X X' + Y Y' + Z Z')$$

and the integral extends over all of the infinite space.

Section 89

Continuation: Two Arbitrary Constant Currents

Equation (5) in section 88 will also remain valid when the first conductor is not linear. For we can conceive of every closed, nonlinear current as a system of linear currents. ΣX, ΣY, ΣZ, respectively, can then be substituted for X, Y, Z and ΣP for P. After this is done, one can again denote the sums with simple letters so that formula (5) will occur once again.

In the same manner, one can also take the two currents to be nonlinear. Equation (5) in section 88 will then remain valid in its unchanged form. One will then only have to understand that X, Y, Z are the components for the *total* magnetic force that the *non*linear first current exerts and that X′, Y′, Z′ are the components for the total force exerted by the *non*linear second current.

We can now return to equations (1) in section 78. Through them, expression (5) in section 88 can be transformed in such a way that

$$(1)\quad P = \frac{1}{4\pi}\int_{-\infty}^{\infty}\int_{-\infty}^{\infty}\int_{-\infty}^{\infty} dx\,dy\,dz \left\{ \begin{aligned} &\frac{\partial u_2}{\partial z} X' - \frac{\partial u_3}{\partial y} X' \\ &+ \frac{\partial u_3}{\partial x} Y' - \frac{\partial u_1}{\partial z} Y' \\ &+ \frac{\partial u_1}{\partial y} Z' - \frac{\partial u_2}{\partial x} Z' \end{aligned} \right\}.$$

It appears appropriate here to apply integration by parts. We obtain

$$\int \frac{\partial u_2}{\partial z} X'\,dz = u_2 X' - \int u_2 \frac{\partial X'}{\partial z}\,dz.$$

In order to ascertain the definite integral, one has to set up limits for both sides. As a result, the component on the right-hand side that is free of an integral sign will vanish, for u_2, as well as X′, is equal to zero for $z = \pm\infty$. One then obtains

$$\int_{-\infty}^{\infty}\int_{-\infty}^{\infty}\int_{-\infty}^{\infty} \frac{\partial u_2}{\partial z} X'\,dx\,dy\,dz = -\int_{-\infty}^{\infty}\int_{-\infty}^{\infty}\int_{-\infty}^{\infty} u_2 \frac{\partial X'}{\partial z}\,dx\,dy\,dz.$$

The remaining components on the right-hand side of (1) can be transformed in the same way. What then results is

$$P = \frac{1}{4\pi}\int_{-\infty}^{\infty}\int_{-\infty}^{\infty}\int_{-\infty}^{\infty} dx\,dy\,dz \left\{ \begin{aligned} &u_1\left(\frac{\partial Y'}{\partial z} - \frac{\partial Z'}{\partial y}\right) \\ &+ u_2\left(\frac{\partial Z'}{\partial x} - \frac{\partial X'}{\partial z}\right) \\ &+ u_3\left(\frac{\partial X'}{\partial y} - \frac{\partial Y'}{\partial x}\right) \end{aligned} \right\}.$$

The integration can still be simplified. For the differences

$$\frac{\partial Y'}{\partial z} - \frac{\partial Z'}{\partial y},\quad \frac{\partial Z'}{\partial x} - \frac{\partial X'}{\partial z},\quad \frac{\partial X'}{\partial y} - \frac{\partial Y'}{\partial x}$$

are only different from zero in the interior of the second conductor[section 66, equation (2); section 77, equations (1), (2), and (3)]. If we then denote a spatial element of the second conductor as dS′, we now obtain

$$(2)\quad P = \frac{1}{4\pi}\int dS' \left\{ \begin{array}{l} u_1\left(\frac{\partial Y'}{\partial z} - \frac{\partial Z'}{\partial y}\right) \\ + u_2\left(\frac{\partial Z'}{\partial x} - \frac{\partial X'}{\partial z}\right) \\ + u_3\left(\frac{\partial X'}{\partial y} - \frac{\partial Y'}{\partial x}\right) \end{array} \right\},$$

and the integration only extends over the space of the second conductor.

Functions u_1, u_2, u_3 are expressed by equations (5) in section 78. For,

$$(3)\quad u_1 = -\int \frac{i_1\, dS}{r},\quad u_2 = -\int \frac{i_2\, dS}{r},\quad u_3 = -\int \frac{i_3\, dS}{r}.$$

Here, u_1, u_2, u_3 only refer to the magnetic forces which are exerted from the first current. Therefore, in equations (3), one has to understand dS as being a spatial element in the interior of the first conductor; i_1, i_2, i_3 are the components of the specific current intensity in one point of this spatial element; and r is the distance of this same point from point (x, y, z). The integrations in (3) will only extend over the space in the first conductor.

Furthermore, if we denote i'_1, i'_2, i'_3 as the components of the specific current intensity in a point in the second conductor, then what we have according to equations (1), (2), and (3) in section 77 is

$$(4)\quad \begin{aligned} 4\pi i'_1 &= \frac{\partial Z'}{\partial y} - \frac{\partial Y'}{\partial z}, \\ 4\pi i'_2 &= \frac{\partial X'}{\partial z} - \frac{\partial Z'}{\partial x}, \\ 4\pi i'_3 &= \frac{\partial Y'}{\partial x} - \frac{\partial X'}{\partial y}. \end{aligned}$$

We will first of all introduce these into equation (2) and get

$$(5)\quad P = -\int dS' (u_1 i'_1 + u_2 i'_2 + u_3 i'_3).$$

The integration is to be extended over the space of the entire second conductor.

In the same way, one also could have obtained the expression

$$(6)\quad P = -\int dS (u'_1 i_1 + u'_2 i_2 + u'_3 i_3).$$

Here, u'_1, u'_2, u'_3 are associated with the magnetic forces that the second current exerts, and i_1, i_2, i_3 are the components of the specific current intensity in a point in the interior of the first conductor. One has to extend the integration in (6) over the space of the first conductor.

If one now substitutes u_1, u_2, u_3 in equation (5) by their expressions from (3), then we get

$$P = \int dS' \left\{ i'_1 \int \frac{i_1\, dS}{r} + i'_2 \int \frac{i_2\, dS}{r} + i'_3 \int \frac{i_3\, dS}{r} \right\}$$

which can be abbreviated

$$(7) \quad P = \int\int \frac{dS\,.\,dS'}{r} \{ i_1\,.\,i'_1 + i_2\,.\,i'_2 + i_3\,.\,i'_3 \}.$$

It is easy to recognize the significance of the expression, $i_1 \cdot i'_1 + i_2 \cdot i'_2 + i_3 \cdot i'_3$, for, when i denotes the total specific current intensity in a point of dS, then, according to equation (9) in section 54, $i_1 = i \cos a$, $i_2 = i \cos b$, $i_3 = i \cos c$. And what we will correspondingly obtain is $i'_1 = i' \cos a'$, $i'_2 = i' \cos b'$, $i'_3 = i' \cos c'$, when i' signifies the total specific current intensity in a point of dS'. Therefore

$$i_1\,.\,i'_1 + i_2\,.\,i'_2 + i_3\,.\,i'_3 = i\,.\,i' (\cos a \ \cos a' + \cos b \ \cos b' + \cos c \ \cos c') = i\,.\,i' \cos (i\, i'),$$

and equation (7) finally changes into the following:

$$(8) \quad P = \int\int \frac{dS\,.\,dS'}{r}\,.\, i\,.\,i' \cos (i\, i').$$

Section 90

Continuation: Two Linear Constant Currents

We will now return to the special case of two constant, closed linear currents. The first conductor [Figure 48] is a wire loop of cross-section q. The axis in the interior of the wire running perpendicular to all cross-sections is a curve, whose length from a fixed point to point (x, y, z) shall be denoted as s. Likewise, the second conductor is also a loop. Its cross-section will be denoted as q'. We will select a fixed origin on its axis and a movable point (x', y', z'). The arc of the axis lying between both points will have the length s'.

In comparison to the lengths of the wires, cross-sections q and q' will

Figure 48

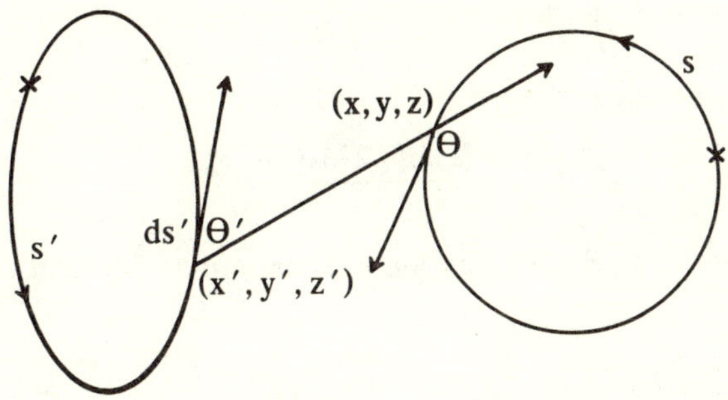

be so small that in all points on one and the same cross-section, the specific current intensity will be constant and directed perpendicular to the cross-section everywhere. We will denote the specific current intensity in point (x, y, z) of the first conductor as i and in point (x′, y′, z′) of the second conductor as i′.

Hereafter, x, y, z, q, and i are functions of s, and x′, y′, z′, q′ and i′ are functions of s′. Section 61 will hold for constant linear currents. So, qi=J, q′i′=J′, and J and J′ are constant here.

In previous cases, dS=dq ds and dS′=dq′ds′. Consequently, equation (8) in section 89 will now read

$$P=\int\int\int\int\frac{1}{r}\,dq\,ds\,.\,dq'\,ds'\,.\,i\,.\,i'\,.\cos\,(i\,i')$$

or, in a more condensed form,

$$(1)\quad P=J\,J'\int\int\frac{ds\,ds'}{r}\cos\,(i\,i').$$

Angle (ii′) will be the same angle that arc elements ds and ds′ form with each other. Therefore, we will have

$$(2)\quad \cos\,(i\,i')=\frac{\partial x}{\partial s}\frac{\partial x'}{\partial s'}+\frac{\partial y}{\partial s}\frac{\partial y'}{\partial s'}+\frac{\partial z}{\partial s}\frac{\partial z'}{\partial s'}.$$

Furthermore, it is to be noted that $r^2=(x-x')^2+(y-y')^2+(z-z')^2$, from which, it turns out,

$$(3)\quad \frac{1}{2}\frac{\partial^2(r^2)}{\partial s\,\partial s'}=\frac{\partial\left(r\frac{\partial r}{\partial s}\right)}{\partial s'}=-\left(\frac{\partial x}{\partial s}\frac{\partial x'}{\partial s'}+\frac{\partial y}{\partial s}\frac{\partial y'}{\partial s'}+\frac{\partial z}{\partial s}\frac{\partial z'}{\partial s'}\right).$$

If one pays attention here, then what can be written instead of equation (1) is

$$(4) \quad P = -JJ' \int\int \frac{ds\,ds'}{r} \cdot \frac{\partial \left(r \frac{\partial r}{\partial s}\right)}{\partial s'}$$

or, what is the same thing,

$$(5) \quad P = -\frac{1}{2} JJ' \int\int \frac{ds\,ds'}{r} \frac{\partial^2 (r^2)}{\partial s\,\partial s'}.$$

One can then transform equation (4). Indefinite integration by parts results in

$$\int \frac{ds'}{r} \frac{\partial \left(r \frac{\partial r}{\partial s}\right)}{\partial s'} = \frac{1}{r} \cdot r \frac{\partial r}{\partial s} - \int \frac{\partial \left(\frac{1}{r}\right)}{\partial s'} r \frac{\partial r}{\partial s} ds'$$

$$= \frac{\partial r}{\partial s} + \int \frac{ds'}{r} \frac{\partial r}{\partial s} \frac{\partial r}{\partial s'}.$$

If one sets up limits, then the section that is free of integral signs drops away because the integration is to be extended along the closed line s′. Consequently, we obtain

$$(6) \quad P = -JJ' \int\int \frac{ds\,ds'}{r} \frac{\partial r}{\partial s} \frac{\partial r}{\partial s'}.$$

Now, by means of differentiation, what results from the expression for r is

$$dr = \frac{x - x'}{r} \left(\frac{\partial x}{\partial s} ds - \frac{\partial x'}{\partial s'} ds'\right)$$
$$+ \frac{y - y'}{r} \left(\frac{\partial y}{\partial s} ds - \frac{\partial y'}{\partial s'} ds'\right)$$
$$+ \frac{z - z'}{r} \left(\frac{\partial z}{\partial s} ds - \frac{\partial z'}{\partial s'} ds'\right),$$

while on the other hand,

$$dr = \frac{\partial r}{\partial s} ds + \frac{\partial r}{\partial s'} ds'.$$

Through comparison we get

$$\frac{x - x'}{r} \cdot \frac{\partial x}{\partial s} + \frac{y - y'}{r} \cdot \frac{\partial y}{\partial s} + \frac{z - z'}{r} \cdot \frac{\partial z}{\partial s} = \frac{\partial r}{\partial s},$$
$$\frac{x - x'}{r} \cdot \frac{\partial x'}{\partial s'} + \frac{y - y'}{r} \cdot \frac{\partial y'}{\partial s'} + \frac{z - z'}{r} \cdot \frac{\partial z'}{\partial s'} = -\frac{\partial r}{\partial s'}.$$

If one then denotes Θ and Θ′ as the angles that the direction of line r forms going from (x′, y′, z′) towards (x, y, z) with the directions of the increasing s and the increasing s′, then one easily recognizes that the two latter equations can also be written as

$$(7)\quad \cos\theta = \frac{\partial r}{\partial s}, \quad \cos\theta' = -\frac{\partial r}{\partial s'}.$$

Consequently, equation (6) will change into the new form

$$(8)\quad P = JJ'\int\int \frac{ds\,ds'}{r}\cos\theta\cos\theta'.$$

Angles Θ and Θ′ are described in Figure 48.

Section 91

Ampère's Law

The force that ds and ds′, the two current elements, exert on each other can be found from function P. We will denote F ds ds′ as the repulsive force that ds and ds′ exert on each other in direction r. If we imagine that during the infinitely small time period dt distance r has changed by δr, then the work performed by the displacement of the elements ds and ds′ is F ds ds′ δr, and the total work in the displacement of both closed conductors is ∫∫ F ds ds′ δr. This total work is equal to the change in P. We thus have the equation

$$(1)\quad \delta P = \int\int F\,ds\,ds'\,\delta r.$$

Now, what can be calculated from equation (5) in section 90 is

$$\delta P = -\frac{1}{2}JJ'\int\int ds\,ds'\left\{-\frac{1}{r^2}\frac{\partial^2(r^2)}{\partial s\,\partial s'}\delta r + \frac{1}{r}\frac{\partial^2\delta(r^2)}{\partial s\,\partial s'}\right\}.$$

The second component on the right-hand side is still to be transformed by means of integration by parts. What results when one integrates indefinitely is

$$\int\frac{1}{r}\frac{\partial^2\delta(r^2)}{\partial s\,\partial s'}ds' = \frac{1}{r}\frac{\partial\delta(r^2)}{\partial s} - \int\frac{\partial\left(\frac{1}{r}\right)}{\partial s'}\frac{\partial\delta(r^2)}{\partial s}ds'.$$

If one then sets up limits, the free part will vanish, because the integration is extended through the closed second conductor. So one gets

$$\int\int \frac{ds\,ds'}{r}\frac{\partial^2 \delta(r^2)}{\partial s\,\partial s'} = -\int\int \frac{\partial\left(\frac{1}{r}\right)}{\partial s'}\frac{\partial \delta(r^2)}{\partial s}\,ds\,ds'.$$

We will transform further. By means of indefinite integration

$$\int \frac{\partial\left(\frac{1}{r}\right)}{\partial s'}\frac{\partial \delta(r^2)}{\partial s}\,ds = \frac{\partial\left(\frac{1}{r}\right)}{\partial s'}\delta(r^2) - \int \delta(r^2)\frac{\partial^2\left(\frac{1}{r}\right)}{\partial s'\,\partial s}\,ds.$$

If one then sets up limits, the free part will vanish once again, for the integration here is extended through the closed first conductor. So one will finally have

$$\int\int \frac{ds\,ds'}{r}\frac{\partial^2 \delta(r^2)}{\partial s\,\partial s'} = \int\int \delta(r^2)\frac{\partial^2\left(\frac{1}{r}\right)}{\partial s\,\partial s'}\,ds\,ds',$$

and therefore,

(2) $$\delta P = -\frac{1}{2}JJ'\int\int \delta r\left\{-\frac{1}{r^2}\frac{\partial^2(r^2)}{\partial s\,\partial s'} + 2r\frac{\partial^2\left(\frac{1}{r}\right)}{\partial s\,\partial s'}\right\}ds\,ds'.$$

Now what we will find by means of differentiation is

$$\frac{\partial\left(\frac{1}{r}\right)}{\partial s} = -\frac{1}{r^2}\frac{\partial r}{\partial s} = -\frac{1}{2r^3}\frac{\partial(r^2)}{\partial s},$$

and, consequently,

$$\frac{\partial^2\left(\frac{1}{r}\right)}{\partial s\,\partial s'} = -\frac{1}{2r^3}\frac{\partial^2(r^2)}{\partial s\,\partial s'} + \frac{3}{2r^4}\frac{\partial r}{\partial s'}\frac{\partial(r^2)}{\partial s}$$
$$= -\frac{1}{2r^3}\frac{\partial^2(r^2)}{\partial s\,\partial s'} + \frac{3}{r^3}\frac{\partial r}{\partial s}\frac{\partial r}{\partial s'}.$$

If we then place this in the expression for δp and consider equations (1) and (2), what results is

(3) $$F\,ds\,ds' = -\frac{JJ'\,ds\,ds'}{r^2}\left(3\frac{\partial r}{\partial s}\frac{\partial r}{\partial s'} - 2\frac{\partial\left(r\frac{\partial r}{\partial s}\right)}{\partial s'}\right).$$

We will denote angle (i i′) as ϵ. Then, with regard to equations (2), (3), and (7) in section 90, equation (3), which was just developed, can also be written

$$(4) \quad F\, ds\, ds' = \frac{J\, J'\, ds\, ds'}{r^2}\,(3 \cos \theta \cos \theta' - 2 \cos \varepsilon).$$

This is the law discovered by Ampère for the electrodynamic interaction of two linear current elements.†

† *Ampère.* Mémoire sur la théorie mathématique des phénomènes électrodynamiques. (Mémoires de l'Académie de Paris. T. VI. 1823.)

SEVENTH DIVISION

Induction

Section 92
The Phenomenon Of Induction

Up to now we have considered the *electromagnetic* interaction between a magnet and a galvanic current, as well as the *electrodynamic* interaction between two galvanic currents. The phenomenon in question is the work performed in all interactions used to move the ponderable carriers of magnetism and galvanic current, respectively. Thus, the magnetic fluids and their ponderable carriers execute collective motion, just as do the galvanic current and its ponderable conductor.

Here, in this case, it is the forces occurring between a magnet and a current, or between two currents, respectively, that effect a change in the relative position of the ponderable carrier. The reverse that is to be expected then is that a change in the mutual position which is imparted to the ponderable carrier can have a new separation of electricities as its consequence, or in other words, stimulate a new current.

Faraday † was the first to experimentally demonstrate that such a *phenomenon* occurs in reality. This kind of current stimulation has been given the name of *induction*.

When an immovable magnet influences the electricity flowing in an immovable conductor, then the positive and the negative electricity experience an equal and equally directed impulse. If one reverses the direction of the current, the magnitude of the impulse remains the same, but its direction is reversed. In both cases, both electricities transfer the common motion impulse that has been imprinted on them to the conductor and this conductor follows the impulse when it is movable.

One can explain this in that the effect that a magnet at rest exerts on one kind of electricity flowing by in a definite direction can be changed into the opposite effect in two ways. The first way is that one keeps the direction constant and substitutes the opposite kind of electricity. The

† *Faraday*. Experimental Researches on Electricity. Series I. II. 1831. 1832.

second occurs when one keeps the kind of electricity constant, but reverses the direction of the current.

If this explanation is correct, then we can predict what will occur when positive and negative electricities flow in opposite directions in a closed conductor and when one then moves the conductor towards the magnet (or the magnet towards the conductor, or both towards each other). Here, we will have a relative change in position which will be of the same direction and magnitude for positive and negative electricity. Consequently, the dissimilarly named electricities will undergo opposite effects. The positive and the negative electricity will be driven from each other in opposite directions: Separation will occur.

Experiments have demonstrated that this result really occurs.

The same results will occur when the magnet is replaced by a constant galvanic current.

Here, it is the conductor's relative motion towards the inducing magnet or the inducing current that brings about the inductive current in that conductor. But induction can also occur with an unchanged mutual position, namely, when the distribution of magnetism undergoes a change in the inducing magnet or the current intensity undergoes a change in the inducing current.

One calls induction by different names, according to whether it is exerted by a magnet or by a galvanic current. In the first case, it is called *magnetic-electric induction* and in the second case, *voltaic induction*.

Section 93

Voltaic Induction – Neumann's Law

We will specifically consider *voltaic induction* for the case where both current conductors are linear. There will be a current with intensity J in one conductor, and with intensity J′ in the other conductor. In order to explore the law of *voltaic induction*, we will first of all investigate the *electrodynamic work* that is performed, using section 90 as our guide. Therefore, we must start with constant currents. We have developed a function P for such constant currents in sections 88 to 91, whose change indicates the electrodynamic work performed in an infinitely small displacement of the current conductor. A linear constant current is one whose current intensity is constant. What we understand as a nonlinear constant current is that kind of current for which the *specific* current intensity is independent of time t every place in the conductor.

We will return to equation (5) in section 90 for two linear constant currents, namely,

(1) $P = -J J' Q.$

Here, we have used Q to denote the factor that is only dependent on the conductor's form and mutual position, namely,

(2) $$Q = \frac{1}{2} \int\int \frac{ds\, ds'}{r} \frac{\partial^2 (r^2)}{\partial s\, \partial s'}.$$

The *electrodynamic* work that is accomplished during the time element from to to t+dt is

(3) $$= -J J' . \frac{dQ}{dt} dt,$$

i.e., equal to the increase that P undergoes in every time element to the extent that J and J′ are taken to be constant. This work is performed by the interaction of the electric particles contained in the current in both of the conductors. But this is not the only work that is performed by virtue of the interaction of both galvanic currents during time element dt, for, as we shall see, there is still the additional *electromotive* work.

To begin with, it is important to note that under certain circumstances the *electrodymanic elementary work* accomplished in the time element from t to t+dt is itself still expressed by (3) when current intensities J and J′ are *not* constant. We will want to take J and J′ as variable, but we will also want to make the assumption that, at all times, the increases in J and J′ that come into existence during the following time element dt are *infinitely small.* Namely, in this case, one can conceive of the increase in J and J′ as though it suddenly came into being in the moment *after* the expiration of the former time element. Then, during the time interval from t to t+dt, current intensities J and J′ are considered as constant, and the *electrodynamic elementary work* performed during the stated time element will be expressed by (3) in actuality.

We will now want to proceed to expand the concept of the potential. Up to now, what we have understood by the potential is a function which is only dependent on the coordinates of the movable particles, whose expression explicitly does not contain time t, and whose difference for the moving particle's starting position and end position indicates the work that is performed in the transfer from the starting position to the end position. So, through the presence of a potential, the work performed is thus only dependent on the starting and end position of the particle and is independent of the paths leading from the starting to the end point. The theory of the conservation of kinetic energy holds in this case.

The concept of the potential should now be expanded, subject to the retention of this function's essential meaning, so that, besides being dependent on the coordinates, it should also be dependent on the velocities of the moving particles. However, as previously, its expression will not explicitly contain time t. In this case, the work that is performed in the transfer from the initial to final position is solely dependent first on the particle's initial and final position, and second on its initial and final velocities. But, it is independent of the paths that are followed and of the velocities attained over this path. Again, the theory of the conservation of kinetic energy applies.

The *electrodynamic* work that is performed by variable current intensities in the time interval from 0 to t is

$$-\int_0^t JJ' \frac{dQ}{dt} dt.$$

This work is composed by the summation of all the elementary work. However, the sum is not only dependent on the particle's initial and final position or on its initial and final velocities. Thus, if *electrodynamic* work were the sole work performed by the interaction of both of the galvanic currents, then, given variable current intensities, no potential would be present and the theory of the conservation of kinetic energy would not be valid.

Now, it has already been emphasized that by virtue of the interaction of both the galvanic currents, *electromotive* work is also accomplished, namely, the work that occurs by means of the induction currents in both conductors. Our actual task is to investigate this work.

We will want to assume that the *total* work that is performed by virtue of the interaction of both currents is constituted in such a way that (in the expanded sense of the word) a potential is present. In other words: We advance the hypothesis that the theory of the conservation of kinetic energy is valid for the motions that are based on both currents' interaction.

Then, when the *electromotive* work that is accomplished in time element dt and that originates in the induction current is added to the work in (3), a sum must result that is a complete differential and, naturally, that is the complete differential of a function that is explicitly only dependent on the coordinates and on the velocities of the moving electric particles.

The desired contribution to expression (3) is

$$(4) \quad J \frac{d(J'Q)}{dt} dt + J' \frac{d(JQ)}{dt} dt.$$

The sum of both expressions (3) and (4) is the complete differential of the function,

(5) $D_1 = J J' Q$,

which is equal (but opposite sign) to function P.

Contribution (4) consists of two parts, of which each can be conceived of as electromotive work. It can be shown easily that the first component expresses the electromotive work in the first conductor and the second component expresses the electromotive work in the second conductor.

For, we are not allowed to forget that work is also performed by the *exterior* electromotive forces and by the forces of free electricity. If we denote E and E′ as the integral values of these electromotive forces for the first and second conductors, respectively, and note that equation (7) in section 61 is valid here, then the *total electromotive* work which is performed in the time interval from t to t + dt turns out to be

$$= J\left\{E + \frac{d(J' Q)}{dt}\right\} dt + J'\left\{E' + \frac{d(JQ)}{dt}\right\} dt.$$

This expression is easy to interpret. It is

$$E + \frac{d(J' Q)}{dt},$$

which is the integral value of the total electromotive force in the first conductor, and

$$E' + \frac{d(J\,Q)}{dt}$$

has the same meaning for the second conductor. The constant currents originate in E and E′. Consequently,

(6) $$\frac{d(J' Q)}{dt}$$

is *the integral value of the electromotive force for the induction current in the first conductor,* and

(7) $$\frac{d(J\,Q)}{dt}$$

is *the integral value of the electromotive force for the induction current in the second conductor.*

This gives the *law of voltaic induction* for two closed linear conductors that was established by *Neumann*.† Experience has confirmed this as being correct.

† *Neumann, F.E.* Allgemeine Gesetze der inducirten elektrischen Ströme. — Ueber ein allgemeines Princip der mathematischen Theorie inducirter elektrischer Ströme. (Abhandlungen der K. Akademie der Wissenschaften zu Berlin. 1845. S.1. 1847. S.1.)

EIGHTH DIVISION

The Fundamental Law Of Electric Interaction

Section 94
The Potential Of The Interaction Of Two Currents

The theorem in section 93 can be transferred immediately to two nonlinear closed currents. One only has to assume that the specific current intensities will undergo only infinitely small changes in time element dt at every place in the first as well as in the second conductor and to advance the hypothesis that the total work that originates in the interaction of both galvanic currents in time element dt is the complete differential of a function which possesses the characteristic properties of a potential (in the broader sense).

In order to comprehend this, one only needs to consider that one can conceive of one as well as the other nonlinear current as each being a system of linear currents.

The train of thought in section 93 will be repeated here. In equations (5), (6), and (7) in section 89, we found these expressions for function P

$$P = -\int dS\,(u_1' i_1 + u_2' i_2 + u_3' i_3)$$

(1) $$= -\int dS'(u_1 i_1' + u_2 i_2' + u_3 i_3')$$

$$= \int\int \frac{dS.dS'}{r}\,(i_1 i_1' + i_2 i_2' + i_3 i_3').$$

We will now also want to consider this function P for the case where the specific current intensity can be independent of time. What matters then are the changes that function P undergoes in time element dt under the various permissible assumptions. $\delta_r P$ will denote the change that occurs when the specific current intensities in both conductors are regarded as independent of t, $\delta_{ri} P$ will denote the change that originates when one considers the specific current intensities in the second conductor only as independent of time t, and $\delta_{ri'}P$ will denote the change that results when the specific current intensitites are assumed to be independent of t only in the first conductor. Finally, dP will be P's complete differential that occurs in time element dt when the mutual position of the elements of the first and second conductor and the specific current intensities every place in both conductors undergo infinitely small changes in every time element.

First of all we have

(2) $dP = -\delta_r P + \delta_{ri} P + \delta_{ri'} P.$

If one presumes that both currents are constant, then, according to section 89, the electrodynamic elementary work performed in the time interval from t to t+dt is

(3) $\delta_r P.$

This expression, then, will still remain correct for the electrodynamic elementary work when the specific current intensities undergo infinitely small changes in time element dt every place in one, as well as in the other conductor. In this case, $\delta_r P$ is not a complete differential and consequently there is no potential present for the electrodynamic work alone. But, *electromotive work* that originates in the interaction of both galvanic currents will still be performed in both conductors.

We will advance the hypothesis that a potential exists for the total work that is performed by virtue of the interaction of both galvanic currents. In order to find this total work, we have to add that kind of contribution that is the sum of a complete differential to equation (3). This contribution is

(4) $-\delta_{ri} P - \delta_{ri'} P$

and the sum is then the complete differential of -P.

Consequently,

$$\begin{aligned} D_1 &= \int dS\,(u_1' i_1 + u_2' i_2 + u_3' i_3) \\ &= \int dS'(u_1 i_1' + u_2 i_2' + u_3 i_3') \qquad (5) \\ &= -\int\int \frac{dS\,.\,dS'}{r}(i_1 i_1' + i_2 i_2' + i_3 i_3') \end{aligned}$$

is the potential of the interaction of both galvanic currents.

The total work breaks down into three terms, namely,

first: the electromotive work in the first conductor:

$$dt \int dS \left(i_1 \frac{du_1'}{dt} + i_2 \frac{du_2'}{dt} + i_3 \frac{du_3'}{dt}\right);$$

second: the electromotive work in the second conductor:

$$dt \int dS' \left(i_1' \frac{du_1}{dt} + i_2' \frac{du_2}{dt} + i_3' \frac{du_3}{dt} \right);$$

third: the electrodynamic work of both currents on one another:

$$dt \int \int dS \, . \, dS' \frac{d\left(\frac{1}{r}\right)}{dt} (i_1 i_1' + i_2 i_2' + i_3 i_3').$$

After we have gotten acquainted with the potential of the interaction of both galvanic currents, we will attempt to explain this interaction from the interaction of the individual electric particles.

For this purpose, it is necessary to discuss generally how the theorems in sections 36 to 43 are to be altered when the potential is not only dependent on the coordinates, but also on the velocities of the moving material points.

Section 95

The Expanded Theorem Of LaGrange:

$$\delta \int_0^t (T - D + S) \, dt = 0$$

We will consider a system of moving material particles. T is the kinetic energy of this system. The expression for the work performed (the potential) at time t may be broken down into two parts, S+D, so that S is only dependent on the particle's coordinates, with D, moreover, still dependent on the velocities. We will denote x, y, z as the coordinates for any one of the material points, and will write (dx/dt)=x′, (dy/dt)=y′, (dz/dt)=z′ as an abbreviation and, correspondingly, the second derivatives. The components of the force acting on point (x, y, z) are X, Y, Z. The work performed in time element dt, after the expiration of time t, is

$$(1) \quad \sum (Xx' + Yy' + Zz') \, dt.$$

The summation is to be extended over all the points. This work is equal to the increase that the potential undergoes in time element dt:

$$(2) \quad \sum (Xx' + Yy' + Zz') \, dt = \left(\frac{dS}{dt} + \frac{dD}{dt} \right) dt.$$

But now we have

$$\frac{dS}{dt} = \sum\left(x'\frac{\partial S}{\partial x} + y'\frac{\partial S}{\partial y} + z'\frac{\partial S}{\partial z}\right), \tag{3}$$

$$\begin{aligned}\frac{dD}{dt} &= \sum\left(x'\frac{\partial D}{\partial x} + y'\frac{\partial D}{\partial y} + z'\frac{\partial D}{\partial z}\right)\\ &+\sum\left(x''\frac{\partial D}{\partial x'} + y''\frac{\partial D}{\partial y'} + z''\frac{\partial D}{\partial z'}\right).\end{aligned} \tag{4}$$

What results from equation (2) is that no term can occur in (dS/dt)+(dD/dt) that does not contain one of the velocity components as a factor. Derivative dS/dt satisfies this condition. In order that the same is the case with dD/dt, no term can be present in D in which the velocities would only occur in the first power. For then, the second component of dD/dt would be loaded with terms which would be free of x′, y′, z′. So, one sees that in D the magnitudes x′, y′, z′ must be at least contained in the *second* power.

As the simplest example, we will take a homogeneous function of the second degree of x′, y′, z′ for D:

$$D = \sum_{ij}\left\{\begin{matrix} A_{ij}\, x_i'\, x_j' + B_{ij}\, y_i'\, y_j' + C_{ij}\, z_i'\, z_j' \\ +2D_{ij}\, z_i'\, x_j' + 2E_{ij}\, x_i'\, y_j' + 2F_{ij}\, y_i'\, z_j' \end{matrix}\right\}. \tag{5}$$

Coefficients $A_{ij} \ldots F_{ij}$ are functions of the coordinates of all the points. Derivative dD/dt will then consist of a homogeneous function of the third degree of x′, y′, z′ and a homogeneous function of the first degree of the same variable, and the coefficients that occur are functions of coordinates x, y, z. However, the homogeneous linear function of x′, y′, z′ that occurs in dD/dt, just like function dS/dt, already has form (1) by itself, and cannot be put into this form in any other way. But, on the other hand, the function of the third degree occurring in dD/dt can be put into form (1) through a manifold of ways. So, the forces in motion are not totally determined by the expression for work.

The theory of the conservation of kinetic energy is expressed in the formula T-S-D=const. We will now inquire how the motion must proceed so that this theory is valid.

We have a clue in section 43 as to how to answer this question. There, it is proven that:

When P is only dependent on the coordinates $q_1, q_2, \ldots$ and this function's expression explicitly does not contain time t and, furthermore, when T is a homogeneous function of the second degree of $q_1', q_2', \ldots,$ then

$$\delta \int_0^t (T + P)\, dt = 0$$

is the necessary and sufficient condition so that T-P=const. S, here, is only a function of coordinates x, y, z . . ., a function whose expression does not explicitly contain time t, while it is T-D that is a homogeneous function of the second degree for x′, y′, z′ . . . . Consequently, we can immediately apply the theorem in section 43, which will now read:

When motion is to proceed so that the theory of the conservation of kinetic energy

$$(6)\quad T - S - D = \text{const.}$$

is valid, then the following necessary and sufficient condition is to be satisfied:

$$(7)\quad \delta \int_0^t (T - D + S)\, dt = 0.$$

This condition yields differential equations of form (6) in section 42. There, one only has to write T-D for T, and S for P to get our case above.

Section 96

The Potential Of Two Electric Particles. Weber's Form

The point is now to apply the theorem in section 95 to the case where the moving material points are electric particles and where the forces that are in motion due to their influence are the forces of mutual attraction and repulsion.

In this problem, D is the potential of the interaction of the electric particles, to the extent that it is co-dependent on the velocities. D will consist of three parts, namely, the potential D_1 of both currents on each other, the potential D_2 of the first current on itself, and the potential D_3 of the second current on itself. According to (5) in section 94,

$$(1)\quad D_1 = -\int\!\!\int \frac{dS\,.\,dS'}{r}(i_1\, i_1' + i_2\, i_2' + i_3\, i_3').$$

When an electrical conductor is moving and the electrical particles inside it are simultaneously in motion, then one can break down the motion of every such particle into two parts, namely, the motion which the conductor imparts to it, and its motion *relative* to the conductor. Then, dx/dt, dy/dt, dz/dt are the components of the absolute velocity of the electrical particle ϵ that is concentrated in point (x, y, z) and $v_1'\, v_2'\, v_3'$ are the components of the absolute velocity of elements of the conductor.

Then,

$$w_1 = \frac{dx}{dt} - v_1,\quad w_2 = \frac{dy}{dt} - v_2,\quad w_3 = \frac{dz}{dt} - v_3$$

are the components of the velocity of the electric particle relative to the conductor.

We will denote x, y, z as the coordinates of a point of dS, and x_1, y_1, z_1 as the coordinates of a point of dS′. Then, $r^2=(x-x_1)^2+(y-y_1)^2+(z-z_1)^2$ and, consequently, by means of differentiation,

$$\frac{\partial^2 (r^2)}{\partial x\, \partial x_1} = \frac{\partial^2 (r^2)}{\partial y\, \partial y_1} = \frac{\partial^2 (r^2)}{\partial z\, \partial z_1} = -2,\quad \frac{\partial^2 (r^2)}{\partial x\, \partial y_1} = \frac{\partial^2 (r^2)}{\partial x_1\, \partial y} = 0,$$

$$\frac{\partial^2 (r^2)}{\partial y\, \partial z_1} = \frac{\partial^2 (r^2)}{\partial y_1\, \partial z} = 0,\quad \frac{\partial^2 (r^2)}{\partial z\, \partial x_1} = \frac{\partial^2 (r^2)}{\partial z_1\, \partial x} = 0.$$

If we now introduce a function F by means of the equation

$$(2)\quad F = i_1'' \frac{\partial (r^2)}{\partial x_1} + i_2'' \frac{\partial (r^2)}{\partial y_1} + i_3'' \frac{\partial (r^2)}{\partial z_1},$$

then we have

$$\frac{\partial F}{\partial x} = -2\, i_1'',\quad \frac{\partial F}{\partial y} = -2\, i_2'',\quad \frac{\partial F}{\partial z} = -2\, i_3''.$$

As a result of this, expression (1) for D_1 can be put into the following form:

$$(3)\quad D_1 = \frac{1}{2}\int\int \frac{dS\, dS'}{r}\left(i_1 \frac{\partial F}{\partial x} + i_2 \frac{\partial F}{\partial y} + i_3 \frac{\partial F}{\partial z}\right).$$

We will begin with integration over the first conductor; thus, with integral

$$\int \frac{dS}{r}\left(i_1 \frac{\partial F}{\partial x} + i_2 \frac{\partial F}{\partial y} + i_3 \frac{\partial F}{\partial z}\right)$$

through integration by parts [equations (1) and (2) in section 20], what we obtain for this is

$$(4)\quad \begin{aligned} &-\int dS \,.\, F \,.\left\{\frac{\partial\left(\frac{i_1}{r}\right)}{\partial x} + \frac{\partial\left(\frac{i_2}{r}\right)}{\partial y} + \frac{\partial\left(\frac{i_3}{r}\right)}{\partial z}\right\} \\ &-\int d\sigma \,.\, \frac{F}{r} \,.\left\{i_1 \frac{\partial x}{\partial n} + i_2 \frac{\partial y}{\partial n} + i_3 \frac{\partial z}{\partial n}\right\}, \end{aligned}$$

and the first of these integrals is to be extended over the space of the first conductor, while the second is to be extended over its surface. However, we will assume that there are currents in which the density of the free electricity does not change in any place [section 57, equation (1)] and from which the conductor's surface is insulated [section 57, equation (2)]. We thus have

$$\frac{\partial i_1}{\partial x}+\frac{\partial i_2}{\partial y}+\frac{\partial i_3}{\partial z}=0,\quad i_1\frac{\partial x}{\partial n}+i_2\frac{\partial y}{\partial n}+i_3\frac{\partial z}{\partial n}=0.$$

Hereafter, the volume integral in (4) simplifies itself and the surface integral drops out completely. As a consequence, expression (3) changes into

$$\text{(5)}\quad D_1=-\frac{1}{2}\iint dS\,dS'.\,F.\left\{i_1\frac{\partial\left(\frac{1}{r}\right)}{\partial x}+i_2\frac{\partial\left(\frac{1}{r}\right)}{\partial y}+i_3\frac{\partial\left(\frac{1}{r}\right)}{\partial z}\right\}.$$

In this formula, one really needs only to work out the differentiation for 1/r and to use function F from equation (2) once again in order to get this new expression,

$$\text{(6)}\quad D_1=\iint\frac{dS\,dS'}{r}\left(i_1\frac{\partial r}{\partial x}+i_2\frac{\partial r}{\partial y}+i_3\frac{\partial r}{\partial z}\right)\left(i'_1\frac{\partial r}{\partial x_1}+i'_2\frac{\partial r}{\partial y_1}+i'_3\frac{\partial r}{\partial z_1}\right).$$

For a further transformation, it will be profitable to take into consideration the connection between the specific current intensities and the velocity components of the single electric particle. For, according to equation (5) in section 54, using the denotation employed here,

$$i_1\,dS=\sum\varepsilon\,w_1=\sum\varepsilon\frac{dx}{dt}-\sum\varepsilon\,v_1,$$

$$i_2\,dS=\sum\varepsilon\,w_2=\sum\varepsilon\frac{dy}{dt}-\sum\varepsilon\,v_2,$$

$$i_3\,dS=\sum\varepsilon\,w_3=\sum\varepsilon\frac{dz}{dt}-\sum\varepsilon\,v_3.$$

The summation extends over all the electric particles contained in spatial element dS. And, for one and the same conductor element, the velocity components v_1, v_2, v_3 can be taken in front of the summation signs. Since free electricity is not present any place in the interior of the conductor, what we have is

$$\text{(7)}\quad \sum\varepsilon=0.$$

Consequently, the last equations simplify themselves and we obtain

$$\text{(8)}\quad i_1\,dS=\sum\varepsilon\frac{dx}{dt},\quad i_2\,dS=\sum\varepsilon\frac{dy}{dt},\quad i_3\,dS=\sum\varepsilon\frac{dz}{dt}.$$

Three corresponding equations result for the spatial element dS′ of the second conductor. With the aid of these equations, expression (6) changes into

$$(9)\quad D_1 = \sum\sum \frac{\varepsilon\varepsilon'}{r}\left(\frac{\partial r}{\partial x}\frac{dx}{dt} + \frac{\partial r}{\partial y}\frac{dy}{dt} + \frac{\partial r}{\partial z}\frac{dz}{dt}\right)$$

$$\times\left(\frac{\partial r}{\partial x_1}\frac{dx_1}{dt} + \frac{\partial r}{\partial y_1}\frac{dy_1}{dt} + \frac{\partial r}{\partial z_1}\frac{dz_1}{dt}\right).$$

The first summation is to be extended over all the electrical particles of the first current and the other summation is to be extended over all the particles of the second current.

Equation (9) can be written even more simply. Namely, if one denotes the change beginning in time dt by r and the change originating in the motion of particle ϵ by δr and the corresponding change in r originating in the motion of particle ϵ' by δ'r, then what finally results is

$$(10)\quad D_1 = \sum\sum \frac{\varepsilon\varepsilon'}{r}\cdot\frac{\delta r}{dt}\cdot\frac{\delta' r}{dt}.$$

This expression gives potential D_1 as dependent on the *absolute* motion of the electric particles. And now, such terms as cancel themselves out in summation can also be added to equation (10); through their introduction it is brought about that only the *relative* velocity occurs.

The sum of these terms is

$$(11)\quad \frac{1}{2}\sum\sum \frac{\varepsilon\varepsilon'}{r}\left\{\left(\frac{\delta r}{dt}\right)^2 + \left(\frac{\delta' r}{dt}\right)^2\right\}.$$

It is easy to see that this double sum has the value of zero. For if we begin in

$$\sum\sum \frac{\varepsilon\varepsilon'}{r}\left(\frac{\delta r}{dt}\right)^2$$

with summation over the second conductor, then factor ϵ can be removed from the inner summation sign. For any single element of the second conductor $1/r(\delta r/dt)^2$ will be constant and $\Sigma\epsilon'=0$. Consequently, every element of the second conductor furnishes a contribution of zero to the sum and, therefore, the whole sum is equal to zero. We can show, in a corresponding manner, that the second component in (11) also has a value of zero.

If we now add contribution (11) to the right-hand side of (10) and write $(\delta r/dt)+(\delta' r/dt)=dr/dt$, then we get

$$(12)\quad D_1 = \frac{1}{2}\sum\sum \frac{\varepsilon\varepsilon'}{r}\left(\frac{dr}{dt}\right)^2.$$

This expression results when one puts:

(13) $D = \frac{1}{2} \frac{\varepsilon \varepsilon'}{r} \left(\frac{dr}{dt}\right)^2$

for the interaction of both single moving particles ϵ and ϵ'.

The electrostatic potential of both particles is

(14) $S = -\frac{\varepsilon \varepsilon'}{r}.$

But, it must be noted here that the quantities of electricity in equations (13) and (14) are measured according to different measures, namely, according to the magnetic one in D and according to the electrostatic one in S. If both expressions are to be combined, they must first be turned into the same kind of measurement. For example, we can introduce an electrostatic measurement into D. This occurs when we write $\epsilon\sqrt{2}/c$ and $\epsilon'\sqrt{2}/c$ instead of ϵ and ϵ' in equations (12) and (13). Magnitude c is a constant which is to be defined by experiment. After this, we will finally obtain the potential of two electrical particles:

(I) $S + D = -\frac{\varepsilon \varepsilon'}{r}\left\{1 - \frac{1}{c^2}\left(\frac{dr}{dt}\right)^2\right\}.$

This expression gives us Weber's Basic Law of interaction between two electrical particles. We will deduce this law in section 97.

Section 97

Weber's Basic Law

We have assumed that the theory of the conservation of kinetic energy is valid in the interaction between electric particles. Consequently, the motion proceeds in such a way that *Lagrange's* expanded theorem (section 95) is satisfied, namely,

(1) $\delta \int_0^t (T - D + S)\, dt = 0.$

We will now take two electric particles which are concentrated in points (x, y, z) and (x_1, y_1, z_1). Their quantities of electricity will be ϵ and ϵ', and their masses m and m_1. In this case,

$$T = \frac{1}{2} m (x'^2 + y'^2 + z'^2) + \frac{1}{2} m_1 (x_1'^2 + y_1'^2 + z_1'^2),$$

$$D = \frac{1}{c^2} \frac{\varepsilon \varepsilon'}{r} \left(\frac{dr}{dt}\right)^2,$$

$$S = -\frac{\varepsilon\,\varepsilon'}{r}.$$

Therefore, we obtain

$$\delta \int_0^t T\,dt = m \int_0^t (x'\,\delta x' + y'\,\delta y' + z'\,\delta z')\,dt$$

$$+ m_1 \int_0^t (x_1'\,\delta x_1' + y_1'\,\delta y_1' + z_1'\,\delta z_1')\,dt.$$

This is the same transformation that was presented in section 39. Therefore, what results is

$$- \delta \int_0^t T\,dt = m \int_0^t \left(\frac{d^2x}{dt^2}\,\delta x + \frac{d^2y}{dt^2}\,\delta y + \frac{d^2z}{dt^2}\,\delta z\right) dt$$

(2)

$$+ m_1 \int_0^t \left(\frac{d^2x_1}{dt^2}\,\delta x_1 + \frac{d^2y_1}{dt^2}\,\delta y_1 + \frac{d^2z_1}{dt^2}\,\delta z_1\right) dt.$$

Furthermore, we have

$$\delta \int_0^t S\,dt = \varepsilon\,\varepsilon' \int_0^t \frac{\delta r}{r^2}\,dt,$$

$$\delta \int_0^t (-D)\,dt = \frac{\varepsilon\,\varepsilon'}{c^2} \int_0^t \frac{\delta r}{r^2} \left(\frac{dr}{dt}\right)^2 dt - 2\,\frac{\varepsilon\,\varepsilon'}{c^2} \int_0^t \frac{1}{r}\,\frac{dr}{dt}\,\frac{d\,\delta r}{dt}\,dt.$$

The last integral can still be transformed. Integration by parts results in

$$\int \frac{1}{r}\,\frac{dr}{dt}\,\frac{d\,\delta r}{dt}\,dt = \frac{1}{r}\,\frac{dr}{dt}\,\delta r - \int \frac{d\left(\frac{1}{r}\,\frac{dr}{dt}\right)}{dt}\,\delta r\,dt.$$

Due to setting up limits, the free part drops away, because $\delta r = 0$ at the beginning and at the end of the motion. Consequently, we obtain

$$\int_0^t \frac{1}{r}\,\frac{dr}{dt}\,\frac{d\,\delta r}{dt}\,dt = -\int_0^t \frac{1}{r}\,\frac{d^2r}{dt^2}\,\delta r\,dt + \int_0^t \frac{1}{r^2} \left(\frac{dr}{dt}\right)^2 \delta r\,dt;$$

and, therefore,

$$\delta \int_0^t (-D + S)\, dt = \int_0^t dt\, \delta r \,.\, \frac{\varepsilon \varepsilon'}{r^2} \tag{3}$$
$$\times \left\{ 1 - \frac{1}{c^2} \left(\frac{dr}{dt}\right)^2 + \frac{2r}{c^2} \frac{d^2 r}{dt^2} \right\}.$$

If we now insert components from (2) and (3) into (1), then what occurs is that *two electric particles ϵ and ϵ' exert a repulsion on each other at distance r, whose direction coincides with their connecting lines and whose magnitude is*

$$\frac{\varepsilon \varepsilon'}{r^2} \left\{ 1 - \frac{1}{c^2} \left(\frac{dr}{dt}\right)^2 + \frac{2r}{c^2} \frac{d^2 r}{dt^2} \right\}.$$

This is *Weber's Basic Law.*†

Section 98

The Potential Of Two Electric Particles. Riemann's Form

We will return to expression (5) in section 94. According to this expression

$$D_1 = -\int\int \frac{dS\, dS'}{r} (i_1 i_1' + i_2 i_2' + i_3 i_3')$$

holds for magnetic measurement, while, on the other hand,

$$D_1 = -\frac{2}{c^2} \int\int \frac{dS\, dS'}{r} (i_1 i_1' + i_2 i_2' + i_3 i_3') \tag{1}$$

holds for electrostatic measurement. If we immediately introduce the velocities here with the aid of equations (8) in section 96 and with the aid of the three corresponding equations for the second conductor, then we obtain

$$D_1 = \frac{1}{c^2} \sum\sum \frac{\varepsilon \varepsilon'}{r} \left\{ -2 \left(\frac{dx}{dt} \frac{dx_1}{dt} + \frac{dy}{dt} \frac{dy_1}{dt} + \frac{dz}{dt} \frac{dz_1}{dt} \right) \right\}. \tag{2}$$

† *Weber.* Elektrodynamische Maassbestimmungen. Theil 1. Seite 99. (Abhandlungen der K. Sächsischen Gesellschaft der Wissenschaften zu Leipzig. 1846.)

We will want to continue to transform in such a way that only the *relative* position and the *relative* motions come into consideration. This is

$$(3)\quad \frac{1}{c^2}\sum\sum\frac{\varepsilon\,\varepsilon'}{r}\left\{\left(\frac{dx}{dt}\right)^2+\left(\frac{dy}{dt}\right)^2+\left(\frac{dz}{dt}\right)^2\right\}=0.$$

Then, we can begin with the summation over the second conductor. The inner summation sign

$$\varepsilon'\left\{\left(\frac{dx}{dt}\right)^2+\left(\frac{dy}{dt}\right)^2+\left(\frac{dz}{dt}\right)^2\right\}$$

will take precedence. For any arbitrary element of the second conductor, $1/r$ is constant. So, $1/r$ can also be taken as a factor in the summation over this element. But $\Sigma\epsilon'=0$ for every single element of the second conductor. Consequently, all single elements of the second conductor will furnish a contribution of zero and, therefore, the whole sum is equal to zero. In a corresponding way, we will show that

$$(4)\quad \frac{1}{c^2}\sum\sum\frac{\varepsilon\,\varepsilon'}{r}\left\{\left(\frac{dx_1}{dt}\right)^2+\left(\frac{dy_1}{dt}\right)^2+\left(\frac{dz_1}{dt}\right)^2\right\}=0.$$

What then results from (2), (3), and (4) is

$$(5)\quad D_1=\frac{1}{c^2}\sum\sum\frac{\varepsilon\,\varepsilon'}{r}\left\{\left(\frac{dx}{dt}-\frac{dx_1}{dt}\right)^2+\left(\frac{dy}{dt}-\frac{dy_1}{dt}\right)^2+\left(\frac{dz}{dt}-\frac{dz_1}{dt}\right)^2\right\}.$$

We will, therefore, assume for two single particles,

$$(\mathrm{II})\quad D=\frac{1}{c^2}\frac{\varepsilon\,\varepsilon'}{r}\left\{\left(\frac{dx}{dt}-\frac{dx_1}{dt}\right)^2+\left(\frac{dy}{dt}-\frac{dy_1}{dt}\right)^2+\left(\frac{dz}{dt}-\frac{dz_1}{dt}\right)^2\right\}.$$

Section 99

Riemann's Basic Law

We will also want to calculate the interaction between two electric particles with the aid of this second expression for D. As in section 97, we will start from the formula

$$(1)\quad \delta\int_0^t (T-D+S)\,dt=0,$$

which expresses *Lagrange's* expanded theorem. So, now, we can follow once more the same path as in section 97. But, it is also permissible to immediately apply formula (6) in section 42, which reads

$$(2)\quad \frac{d\left(\frac{\partial (T-D)}{\partial q'}\right)}{dt} = \frac{\partial (T-D+S)}{\partial q}.$$

Coordinates x, y, z, x_1, y_1, z_1 are to be inserted successively for q. We will carry out the calculation for q=x. It is

$$T = \frac{1}{2} m (x'^2 + y'^2 + z'^2) + \frac{1}{2} m_1 (x_1'^2 + y_1'^2 + z_1'^2),$$

thus,

$$\frac{\partial T}{\partial x'} = m x', \quad \frac{\partial T}{\partial x} = 0.$$

And, therefore, we now have

$$(3)\quad X = m \frac{d^2 x}{dt^2} = \frac{d\left(\frac{\partial D}{\partial x'}\right)}{dt} - \frac{\partial D}{\partial x} + \frac{\partial S}{\partial x}.$$

But, what also results from formula (II) in section 98 is

$$\frac{\partial D}{\partial x'} = 2 \frac{\varepsilon \varepsilon'}{c^2} \frac{1}{r} \left(\frac{dx}{dt} - \frac{dx_1}{dt}\right),$$

$$\frac{\partial D}{\partial x} = - \frac{\varepsilon \varepsilon'}{c^2} \frac{1}{r^2} \frac{\partial r}{\partial x} \left\{\left(\frac{dx}{dt} - \frac{dx_1}{dt}\right)^2 + \left(\frac{dy}{dt} - \frac{dy_1}{dt}\right)^2 + \left(\frac{dz}{dt} - \frac{dz_1}{dt}\right)^2\right\}.$$

And, finally,

$$\frac{\partial S}{\partial x} = \frac{\varepsilon \varepsilon'}{r^2} \frac{\partial r}{\partial x}.$$

When one inserts this into equation (3), one obtains

$$(4)\quad \begin{aligned} X = {} & \frac{\varepsilon \varepsilon'}{r^2} \frac{\partial r}{\partial x} + \frac{\varepsilon \varepsilon'}{c^2} \frac{d\left\{\frac{2}{r}\left(\frac{dx}{dt} - \frac{dx_1}{dt}\right)\right\}}{dt} \\ & + \frac{\varepsilon \varepsilon'}{c^2} \frac{1}{r^2} \frac{\partial r}{\partial x} \left\{\left(\frac{dx}{dt} - \frac{dx_1}{dt}\right)^2 + \left(\frac{dy}{dt} - \frac{dy_1}{dt}\right)^2 + \left(\frac{dz}{dt} - \frac{dz_1}{dt}\right)^2\right\}. \end{aligned}$$

just as

$$(5)\quad \begin{aligned} Y = {} & \frac{\varepsilon \varepsilon'}{r^2} \frac{\partial r}{\partial y} + \frac{\varepsilon \varepsilon'}{c^2} \frac{d\left\{\frac{2}{r}\left(\frac{dy}{dt} - \frac{dy_1}{dt}\right)\right\}}{dt} \\ & + \frac{\varepsilon \varepsilon'}{c^2} \frac{1}{r^2} \frac{\partial r}{\partial y} \left\{\left(\frac{dx}{dt} - \frac{dx_1}{dt}\right)^2 + \left(\frac{dy}{dt} - \frac{dy_1}{dt}\right)^2 + \left(\frac{dz}{dt} - \frac{dz_1}{dt}\right)^2\right\}, \end{aligned}$$

$$Z = \frac{\varepsilon\,\varepsilon'}{r^2}\frac{\partial r}{\partial z} + \frac{\varepsilon\,\varepsilon'}{c^2}\frac{d\left\{\frac{2}{r}\left(\frac{dz}{dt}-\frac{dz_1}{dt}\right)\right\}}{dt} \tag{6}$$
$$+\frac{\varepsilon\,\varepsilon'}{c^2}\frac{1}{r^2}\frac{\partial r}{\partial z}\left\{\left(\frac{dx}{dt}-\frac{dx_1}{dt}\right)^2+\left(\frac{dy}{dt}-\frac{dy_1}{dt}\right)^2+\left(\frac{dz}{dt}-\frac{dz_1}{dt}\right)^2\right\}.$$

Tests have not yet been conducted successfully for moving free electricity.

Section 100

The Effect Of All The Particles ϵ On A Particle ϵ'. Riemann's Law

In order to investigate the effect of all the electric particles ϵ' on one particle ϵ, we have to put

$$S = \varepsilon\sum\left(-\frac{\varepsilon'}{r}\right) = \varepsilon\, V, \tag{1}$$

where V denotes the electrostatic potential function of particle ϵ' at point (x, y, z). Regarding D, we have to distinguish between the two hypotheses (sections 96 and 99). According to *Weber's* formula

$$D = \varepsilon\sum\frac{\varepsilon'}{c^2}\frac{1}{r}\left(\frac{dr}{dt}\right)^2; \tag{2a}$$

while, on the other hand, according to *Riemann's* formula,

$$D = \varepsilon\sum\frac{\varepsilon'}{c^2}\frac{1}{r}\left\{\left(\frac{dx}{dt}-\frac{dx_1}{dt}\right)^2+\left(\frac{dy}{dt}-\frac{dy_1}{dt}\right)^2+\left(\frac{dz}{dt}-\frac{dz_1}{dt}\right)^2\right\}. \tag{2b}$$

We want to deal first with the latter formula. If the square in (2a) is calculated, then D breaks down into three components, namely,

$$D = \varepsilon\sum\frac{\varepsilon'}{c^2}\frac{1}{r}\left\{\left(\frac{dx}{dt}\right)^2+\left(\frac{dy}{dt}\right)^2+\left(\frac{dz}{dt}\right)^2\right\}$$
$$+\varepsilon\sum\frac{\varepsilon'}{c^2}\frac{1}{r}\left\{\left(\frac{dx_1}{dt}\right)^2+\left(\frac{dy_1}{dt}\right)^2+\left(\frac{dz_1}{dt}\right)^2\right\}$$
$$-2\varepsilon\sum\frac{\varepsilon'}{c^2}\frac{1}{r}\left\{\frac{dx}{dt}\frac{dx_1}{dt}+\frac{dy}{dt}\frac{dy_1}{dt}+\frac{dz}{dt}\frac{dz_1}{dt}\right\}.$$

If we denote the velocity of particle ϵ by v, and the velocity of particle ϵ' by v', then this can be written in a shorter form:

$$D = \frac{\varepsilon}{c^2} v^2 \sum \frac{\varepsilon'}{r} + \frac{\varepsilon}{c^2} \sum \frac{\varepsilon'}{r} v'^2$$
$$- 2 \frac{\varepsilon}{c^2} \sum \frac{\varepsilon'}{r} \left\{ \frac{dx}{dt} \frac{dx_1}{dt} + \frac{dy}{dt} \frac{dy_1}{dt} + \frac{dz}{dt} \frac{dz_1}{dt} \right\}$$
$$= - \frac{\varepsilon}{c^2} v^2 V + \frac{\varepsilon}{c^2} \sum \frac{\varepsilon'}{r} v'^2 - 2 \frac{\varepsilon}{c^2} \frac{dx}{dt} \sum \frac{\varepsilon'}{r} \frac{dx_1}{dt}$$
$$- 2 \frac{\varepsilon}{c^2} \frac{dy}{dt} \sum \frac{\varepsilon'}{r} \frac{dy_1}{dt} - 2 \frac{\varepsilon}{c^2} \frac{dz}{dt} \sum \frac{\varepsilon'}{r} \frac{dz_1}{dt},$$

and we will want to set up the abbreviation

$$\sum \frac{\varepsilon'}{r} v'^2 = W, \quad \sum \frac{\varepsilon'}{r} \frac{dx_1}{dt} = u_1,$$
$$\sum \frac{\varepsilon'}{r} \frac{dy_1}{dt} = u_2, \quad \sum \frac{\varepsilon'}{r} \frac{dz_1}{dt} = u_3.$$

Then, we will have

$$\text{(3)} \quad D = - \frac{\varepsilon}{c^2} v^2 V + \frac{\varepsilon}{c^2} W - 2 \frac{\varepsilon}{c^2} \left(u_1 \frac{dx}{dt} + u_2 \frac{dy}{dt} + u_3 \frac{dz}{dt} \right).$$

Functions V, W, u_1, u_2, u_3 satisfy *Laplace's* equation and, consequently, D does too, to the extent that it is dependent on x,y,z:

$$\text{(4)} \quad \frac{\partial^2 D}{\partial x^2} + \frac{\partial^2 D}{\partial y^2} + \frac{\partial^2 D}{\partial z^2} = 0.$$

We still want to establish the change of V, which occurs in time element dt, so that the particles ϵ' are in motion and x,y,z are taken as constant. What results is

$$\frac{\partial V}{\partial t} = - \sum \varepsilon' \frac{\partial \left(\frac{1}{r} \right)}{\partial x_1} \frac{dx_1}{dt} - \sum \varepsilon' \frac{\partial \left(\frac{1}{r} \right)}{\partial y_1} \frac{dy_1}{dt} - \sum \varepsilon' \frac{\partial \left(\frac{1}{r} \right)}{\partial z_1} \frac{dz_1}{dt}.$$

Now, however, we have

$$\frac{\partial \left(\frac{1}{r} \right)}{\partial x_1} = - \frac{\partial \left(\frac{1}{r} \right)}{\partial x},$$

consequently,

$$- \sum \varepsilon' \frac{\partial \left(\frac{1}{r} \right)}{\partial x_1} \frac{dx_1}{dt} = \sum \varepsilon' \frac{\partial \left(\frac{1}{r} \right)}{\partial x} \frac{dx_1}{dt} = \frac{\partial u_1}{\partial x},$$

just as

$$-\sum \varepsilon' \frac{\partial\left(\frac{1}{r}\right)}{\partial y_1} \frac{dy_1}{dt} = \sum \varepsilon' \frac{\partial\left(\frac{1}{r}\right)}{\partial y} \frac{dy_1}{dt} = \frac{\partial u_2}{\partial y},$$

$$-\sum \varepsilon' \frac{\partial\left(\frac{1}{r}\right)}{\partial z_1} \frac{dz_1}{dt} = \sum \varepsilon' \frac{\partial\left(\frac{1}{r}\right)}{\partial z} \frac{dz_1}{dt} = \frac{\partial u_3}{\partial z}.$$

Therefore the expression for $\partial V/\partial t$ changes into the following:

$$(5) \quad \frac{\partial V}{\partial t} = \frac{\partial u_1}{\partial x} + \frac{\partial u_2}{\partial y} + \frac{\partial u_3}{\partial z}.$$

On the basis of this differential equation, one can make an assumption about the meaning of functions V, u_1, u_2, u_3. One can assume that the electric effect is mediated through an aether. By virtue of equation (5), V can then be considered the density and u_1, u_2, u_3 the current intensities of this aether.

Section 101

Continuation: Weber's Law

We will also want to establish the potential for the effect of all the particles ϵ' on one particle ϵ according to *Weber's* theory.

First of all, we again have

$$(1) \quad S = \varepsilon V.$$

This function satisfies *Laplace's* equation. As an abbreviation, the sum of the three derivatives can be used for any function F:

$$\frac{\partial^2 F}{\partial x^2} + \frac{\partial^2 F}{\partial y^2} + \frac{\partial^2 F}{\partial z^2} = \Delta_2 F.$$

Through this designation, we obtain

$$(2) \quad \Delta_2 S = 0.$$

Function D can now be taken out of equation (2a) in section 100. But, since $r^2=(x-x_1)^2+(y-y_1)^2+(z-z_1)^2$, consequently,

$$\frac{dr}{dt} = \frac{(x-x_1)}{r}\left(\frac{dx}{dt} - \frac{dx_1}{dt}\right) + \frac{(y-y_1)}{r}\left(\frac{dy}{dt} - \frac{dy_1}{dt}\right)$$
$$+ \frac{(z-z_1)}{r}\left(\frac{dz}{dt} - \frac{dz_1}{dt}\right).$$

We will insert this into the expression for D, getting

$$D=\frac{\varepsilon}{c^2}\sum\frac{\varepsilon'}{r^3}\Big\{(x-x_1)\Big(\frac{dx}{dt}-\frac{dx_1}{dt}\Big)+(y-y_1)\Big(\frac{dy}{dt}-\frac{dy_1}{dt}\Big) \tag{3}$$
$$+(z-z_1)\Big(\frac{dz}{dt}-\frac{dz_1}{dt}\Big)\Big\}^2.$$

To the extent that this function is dependent on x,y,z, it does not satisfy *Laplace's* equation, but the complicated differential equation

$$\Delta_2\,\Delta_2\,D=0. \tag{4}$$

In order to prove this, we have

$$\frac{1}{r^3}\Big\{(x-x_1)\Big(\frac{dx}{dt}-\frac{dx_1}{dt}\Big)+(y-y_1)\Big(\frac{dy}{dt}-\frac{dy_1}{dt}\Big)$$
$$+(z-z_1)\Big(\frac{dz}{dt}-\frac{dz_1}{dt}\Big)\Big\}=G,$$

$$(x-x_1)\Big(\frac{dx}{dt}-\frac{dx_1}{dt}\Big)+(y-y_2)\Big(\frac{dy}{dt}-\frac{dy_1}{dt}\Big)$$
$$+(z-z_1)\Big(\frac{dz}{dt}-\frac{dz_1}{dt}\Big)=H.$$

The single summands in D, then, regardless of constant factors, are of the form G·H. But,

$$\Delta_2\,(G\,H)=G\,\Delta_2\,H+H\,\Delta_2\,G$$
$$+2\Big(\frac{\partial G}{\partial x}\,\frac{\partial H}{\partial x}+\frac{\partial G}{\partial y}\,\frac{\partial H}{\partial y}+\frac{\partial G}{\partial z}\,\frac{\partial H}{\partial z}\Big),$$

and it can easily be proven through differentiation that $\Delta_2 G=0$, $\Delta_2 H=0$. Consequently, we will obtain the simpler equation

$$\Delta_2\,(G\,H)=2\Big(\frac{\partial G}{\partial x}\,\frac{\partial H}{\partial x}+\frac{\partial G}{\partial y}\,\frac{\partial H}{\partial y}+\frac{\partial G}{\partial z}\,\frac{\partial H}{\partial z}\Big).$$

Factors

$$\frac{\partial H}{\partial x},\ \frac{\partial H}{\partial y},\ \frac{\partial H}{\partial z}$$

will be independent of x,y,z. So,

$$\Delta_2\,\Delta_2\,(GH)=2\Big(\frac{\partial H}{\partial x}\,\frac{\partial\Delta_2 G}{\partial x}+\frac{\partial H}{\partial y}\,\frac{\partial\Delta_2 G}{\partial y}+\frac{\partial H}{\partial z}\,\frac{\partial\Delta_2 G}{\partial z}\Big)$$

and this is equal to zero because $\Delta_2 G=0$. This also proves equation (4).

Thus, *Weber's* hypothesis, in the case of the problem at hand, leads to a more complicated differential equation.

Section 102
The Motion Of Particle ϵ. Riemann's Law

We now want to deduce the equations of motion for particle ϵ. First, according to *Riemann's* hypothesis

(1) $S = \varepsilon V.$

(2) $D = -\frac{\varepsilon}{c^2} v^2 V + \frac{\varepsilon}{c^2} W - 2\frac{\varepsilon}{c^2}\left(u_1 \frac{dx}{dt} + u_2 \frac{dy}{dt} + u_3 \frac{dz}{dt}\right).$

Lagrange's expanded theorem will be valid for the motion and what results from it is like what resulted in section 90, equation (2):

$$\frac{d\left(\frac{\partial (T-D)}{\partial q'}\right)}{dt} = \frac{\partial (T-D+S)}{\partial q}.$$

Coordinates x, y, z are to be successively inserted here for q. We will obtain results for q=x in the same manner as in section 99, equation (3):

(3) $$m\frac{d^2x}{dt^2} = \frac{d\left(\frac{\partial D}{\partial x'}\right)}{dt} - \frac{\partial D}{\partial x} + \frac{\partial S}{\partial x}.$$

The partial derivatives $\partial D/\partial x$ and $\partial S/\partial x$ which are taken with respect to x are independent of the acceleration. But, of course, acceleration occurs in

$$\frac{d\left(\frac{\partial D}{\partial x'}\right)}{dt}.$$

Namely,

$$\frac{\partial D}{\partial x'} = -2\frac{\varepsilon}{c^2} V\frac{dx}{dt} - 2\frac{\varepsilon}{c^2} u_1.$$

Consequently,

$$\frac{d\left(\frac{\partial D}{\partial x'}\right)}{dt} = -2\frac{\varepsilon}{c^2} V\frac{d^2x}{dt^2} - 2\frac{\varepsilon}{c^2}\frac{dV}{dt}\frac{dx}{dt} - 2\frac{\varepsilon}{c^2}\frac{du_1}{dt},$$

or, even more briefly,

$$\frac{d\left(\frac{\partial D}{\partial x'}\right)}{dt} = -2\,\frac{\varepsilon}{c^2}\,V\,\frac{d^2x}{dt^2} + \frac{\delta\left(\frac{\partial D}{\partial x'}\right)}{dt},$$

when one indicates by δ a differentiation by t, in which dx/dt is considered as constant. If one introduces this into equation (3), then the result is

$$(4)\quad \left(m + \frac{2\,\varepsilon}{c^2}\,V\right)\frac{d^2x}{dt^2} = \frac{\delta\left(\frac{\partial D}{\partial x'}\right)}{dt} - \frac{\partial D}{\partial x} + \frac{\partial S}{\partial x}.$$

We will obtain both of the other equations in the same way:

$$(5)\quad \left(m + \frac{2\,\varepsilon}{c^2}\,V\right)\frac{d^2y}{dt^2} = \frac{\delta\left(\frac{\partial D}{\partial y'}\right)}{dt} - \frac{\partial D}{\partial y} + \frac{\partial S}{\partial y},$$

$$(6)\quad \left(m + \frac{2\,\varepsilon}{c^2}\,V\right)\frac{d^2z}{dt^2} = \frac{\delta\left(\frac{\partial D}{\partial z'}\right)}{dt} - \frac{\partial D}{\partial z} + \frac{\partial S}{\partial z}.$$

Section 103

Continuation: Weber's Law

The equations of motion for *electric* particle ϵ will be finally derived from *Weber's* formula too:

$$(1)\quad S = \varepsilon\,V,$$

$$(2)\quad D = \frac{\varepsilon}{c^2}\sum\frac{\varepsilon'}{r^3}\left\{(x-x_1)\left(\frac{dx}{dt}-\frac{dx_1}{dt}\right) + (y-y_1)\left(\frac{dy}{dt}-\frac{dy_1}{dt}\right) + (z-z_1)\left(\frac{dz}{dt}-\frac{dz_1}{dt}\right)\right\}^2.$$

This results in

$$\frac{\partial D}{\partial x'} = a\,\frac{dx}{dt} + b\,\frac{dy}{dt} + c\,\frac{dz}{dt} + k,$$

in which a, b, c, k are functions of x,y,z that satisfy the partial differential equation $\Delta_2\Delta_2 F = 0$. By means of differentiation by t we obtain

$$\frac{d\left(\frac{\partial D}{\partial x'}\right)}{dt} = a\,\frac{d^2x}{dt^2} + b\,\frac{d^2y}{dt^2} + c\,\frac{d^2z}{dt^2} + g,$$

and function g here is dependent only on coordinates x,y,z and on velocities dx/dt, dy/dt, dz/dt. According to this, the equations for motion read

$$(3)\quad \begin{aligned} (m-a)\frac{d^2x}{dt^2} - b\frac{d^2y}{dt^2} - c\frac{d^2z}{dt^2} &= g - \frac{\partial D}{\partial x} + \frac{\partial S}{\partial x}, \\ -a_1\frac{d^2x}{dt^2} + (m-b_1)\frac{d^2y}{dt^2} - c_1\frac{d^2z}{dt^2} &= g_1 - \frac{\partial D}{\partial y} + \frac{\partial S}{\partial y}, \\ -a_2\frac{d^2x}{dt^2} - b_2\frac{d^2y}{dt^2} + (m-c_2)\frac{d^2z}{dt^2} &= g_2 - \frac{\partial D}{\partial z} + \frac{\partial S}{\partial z}. \end{aligned}$$

So, elimination must take place here first.

Section 104

The Connection With Ampère's Law

Expression (12) in section 96 has been interpreted by us in such a way that the portion of the total potential of two closed currents acting on each other that is dependent on velocities is composed by means of summation of only single potentials. The single potential is generally based on two electric particles ϵ and ϵ'. So, if the question is about the potential D_1 of two currents acting on one another, then one has to combine every particle ϵ of one current with every particle ϵ' of the other current, form the single potential for every such combination, and then sum up all the single potentials. This is how the expression for D_1 in equation (13) in section 96 correctly resulted from equation (12) in the same section and the same holds for the expression for D_1 in equation (5) in section 98, which came from expression (II) in the same section.

If one then uses either

$$(1)\quad S + D = -\frac{\varepsilon\varepsilon'}{r}\left\{1 - \frac{1}{c^2}\left(\frac{dr}{dt}\right)^2\right\}$$

from *Weber's* fundamental law, or

$$(2)\quad \begin{aligned} S+D = -\frac{\varepsilon\varepsilon'}{r} + \frac{\varepsilon\varepsilon'}{c^2 r}\Big\{&\left(\frac{dx}{dt}-\frac{dx_1}{dt}\right)^2 + \left(\frac{dy}{dt}-\frac{dy_1}{dt}\right)^2 \\ &+ \left(\frac{dz}{dt}-\frac{dz_1}{dt}\right)^2\Big\} \end{aligned}$$

from *Riemann's* fundamental law to calculate the total interaction of all electric particles that are generally contained in two closed conductors at rest and in current, then, for every combination of two different particles ϵ and ϵ', one has to set up expressions (1) and (2), respectively, and sum them.

We must distinguish between three different combinations here, namely, two particles at rest, one particle at rest and one moving, and, finally, two moving particles.

We will want to consider the special case of two closed constant currents in order to investigate whether ***Weber's*** basic law, or ***Riemann's*** basic law, repectively, is in agreement with ***Ampère's*** law. The question in ***Ampère's*** law concerns the electrodynamic interaction between two current elements of which one will belong to the first current, while the other will belong to the second current. So, what comes into consideration here is only the interaction between the *moving* electric particles of both constant currents.

It can first of all be proved that the contribution to the total potential of the moving electric particles that orignates in S is equal to zero. For, we can bring all the other ϵ' particles into combination beginning with a single ϵ particle. Then, ϵ will leave the sum sign and the summation of $\Sigma(\epsilon'/r)$ will extend over all of the particles ϵ' different from ϵ. If first we undertake summation over a current element so that $1/r$ can also be placed before the sum sign, then $\Sigma\epsilon'=0$ in every current element for constant current. Thus, all contributions to the sum being formed are zero. This holds for the combination of every single particle ϵ with the particles ϵ' that are different from it. Consequently,

$$(3)\quad -\sum \frac{\varepsilon\,\varepsilon'}{r} = 0.$$

So, only the sum of all the values for D is left for the combinations of each of two moving particles. These combinations will break down into three groups:

First: each single particle of the first current with a single particle of the second current;
Second: two particles each of the first current;
Third: two particles each of the second current.

These groups will successively furnish the potentials which are denoted in section 96 as D_1 D_2 D_3.

D_2 and D_3 are constant for constant currents. If we proceed from (1), then

$$(4)\quad D_2 = \frac{1}{c^2}\sum \frac{\varepsilon\,\varepsilon'}{r}\left(\frac{dr}{dt}\right)^2,$$

when the summation is extended over all the combinations of the particles of the first current.

Because the conductor is assumed to have an invariant form, we can then base a *fixed, connected* coordinate system x,y,z on the same conductor. Then, given a *constant* current, function

$$\frac{\varepsilon\,\varepsilon'}{r}\left(\frac{dr}{dt}\right)^2$$

is only dependent on x,y,z on the one hand and on x_1, y_1, z_1 on the other hand. If one next takes a single ϵ and then sums for all ϵ', then the sum is uniquely and solely a function of x,y,z, i.e., of the coordinates of every particle ϵ. But, if one forms these sums for every value-combination x,y,z that belongs to points in the interior of the conductor and then combines all of these sums together through addition, the result is constant.

The same holds for sum

$$(5)\quad D_2 = \frac{1}{c^2}\sum\frac{\varepsilon\,\varepsilon'}{r}\left\{\left(\frac{dx}{dt}-\frac{dx_1}{dt}\right)^2+\left(\frac{dy}{dt}-\frac{dy_1}{dt}\right)^2+\left(\frac{dz}{dt}-\frac{dz_1}{dt}\right)^2\right\},$$

when it is extended over all combinations of particles of the first current.

We can prove that D_3 is constant with constant currents in the same way.

Thus, with constant currents, the total work performed by the moving electric particles is equal to the change in D_1 alone. So, according to this, it turns out that *Weber* and *Riemann's* basic laws are in agreement with *Ampère's* because *Ampère's* law is on constant currents. In his observations, *Ampère* watched for the equilibrium position of moving current conductors which have constant current flowing through them. It was from these observations that he abstracted his law. Because we have now deduced *Ampère's* law from D_1 and we have been able to produce the expression for D_1 from *Weber's* and also from *Riemann's* basic laws, then their complete agreement has been proven in fact.†

† *Kirchhoff* published two treatises on the movement of electricity in wire-shaped and arbitrary conductors, both in *Poggendorff's* Annalen, Bd. 100 (S.193) and Bd. 102 (S.529). Here, the electromotive force is viewed as originating in the free electricity which is present and in the induction which occurs as a consequence of the changes in the current intensity in all segments of the conductor. As a result of this, *Kirchhoff* gets currents in which it is only as an exception that the density of the free electricity in the interior of the conductor is equal to zero. These investigations by *Kirchhoff* form the starting point for the developments reported by *Weingarten* and *Lorberg*. (*Weingarten*. Ueber die Bewegung der Elektricität in Leitern. *Borchardt's* Journal. Bd. 63 — *Lorberg*. Zur Theorie der Bewegung der Elektricitat in nicht linearen Leitern. *Borchardt's* Journal Bd. 71. S. 53.)

In 1858, *Riemann* presented a treatise to the Royal Society of Sciences at Göttingen, but later withdrew it. This same treatise is printed in *Poggendorf's* Annalen, Bd. 131 (S. 237) under the title, "Ein Betrag zur Elektrodynamik." This states the hypothesis that the force present in an electric particle at time t first begins its effect at a finite distance on another such paritcle at a later time t + Δ t. We can also find this fundamental idea in a contemporary (1867) treatise published by *L. Lorenz:* Ueber die Identität der Schwingungen des Lichts mit den elektrischen Strömen, (*Poggendorff's* Annalen Bd. 131. S. 243). *C. Neumann* dealt with the same fundamental ideas further (Die Principien der Elektrodynamik. Tübingen 1868. Gratulationsschrift. — Allgemeine Betrachtungen über das *Weber*'sche Gesetz. Mathematische Annalen Bd. 8. 1875.)

In recent years, *Weber's* basic law has become the subject of a controversy incited by *Helmholtz.* One should see these treatises about it:

Helmholtz: Ueber die Bewegungsgleichungen für ruhende leitende Korper. (*Borchardt's* Journal Bd.72) — Ueber die Theorie der Elektrodynamik. (*Borchardt's* Journal Bd. 75 and Bd. 78.)

Weber: Elektrodynamische Maassbestimmungen, insbesondere über das Princip der Erhaltung der Energie. (Abhandlungen der mathematische-physischen Klasse der K. Sächsischen Gesellschaft der Wissenschaften Bd. 10)

C. Neumann: Ueber die den Kräften elektrodynamischen Ursprungs zuzuschreibenden Elementargesetze. (Ibid., Bd. 10)

Also: *C. Neuman's* papers in volumes 5 and 6 of the *Mathematischen Annalen* and his monograph: Die elektrischen Kräfte. Theil 1. Leipzig 1873.

Both of *H. Weber's* papers are of particular interest for the mathematician: Ueber die *Bessel*'schen Functionen und ihre Anwendung auf die Theorie der elektrischen Ströme. (*Borchardt's* Journal Bd. 75) — Ueber die stationären Strömungen der Elektricität in Cylindern. (*Borchardt's* Journal Bd. 76)

These texts deserve mention:

Beer: Einleitung in die Elektrostatik, die Lehre vom Magnetismus und die Elektrodynamik. Braunschweig 1865.

Wiedemann: Die Lehre vom Galvanismus und Elektromagnetismus. Second edition. Bd. I, II, 1 and 2. Braunschweig 1872. 1873. 1874.

Maxwell: A treatise on electricity and magnetism. Vol. I.II. Oxford 1873.

One will find a detailed overview of the literature in *Wiedemann's* book.

A Contribution To Electrodynamics

(From Poggendorf's *Annalen der Physik und Chemie,* Bd. CXXXI)

I have taken the liberty of informing the Royal Society of an observation which brings the theory of electricity and of magnetism into close coherence with the theories of light and radiating heat. I have found that the electrodynamic effects of galvanic current can be explained if one assumes that the effect of an electric mass on other masses does not occur instantaneously, but is propagated with a constant velocity (equal to the velocity of light inside the boundaries of observational error). In this assumption, the differential equation for the propagation of electricity becomes the same as the equation for the propagation of light and radiating heat.

Let S and S′ be two conductors through which constant galvanic currents flow and which do not move in relation to each other. ϵ is a particle of electric mass inside conductor S which is located at point (x, y, z) at time t. ϵ' is a particle of electric mass of S′ which is located at point (x′, y′, z′) at time t′. In terms of the motion of the particles of electric mass, in which the positive and negative electric particles have opposite directions in every particle of the conductor, I will make the assumption that the particles of electric mass are distributed at every instant in such a way that the sums, $\Sigma\epsilon f(x,y,z)$ and $\Sigma\epsilon' f(x',y',z')$, which are extended over all the mass particles in the conductor, can be neglected in favor of those sums that are extended only over the positive or only over the negative particles of electric mass as long as function f and its differential equation are continuous.

This provision can be satisfied in many diverse ways. If one assumes, for example, that the conductor is crystalline in its smallest particles so that the same relative distribution of electricity is periodically repeated at distances which are definite and infinitely small compared to the conductor's dimensions, and if β designates the length of such a period, then these sums are infinitely small like $c\beta^n$ — if f and its derivatives are continuous up to the (n-1)th order — and are infinitely small like e-(c/β) — if they are all continuous.

The Empirical Law of Electrodynamic Effects

If the specific current intensities, according to mechanical measurement at time t in points (x, y, z), are parallel to the three axes u, v, w, and if, in points (x′, y′, z′), they are parallel to u′, v′, w′, and if r describes the distance between both points, while c is the constant determined by *Kohlrausch* and *Weber*, then, according to experiments, the potential of the forces exerted by S and S′ is

$$-\frac{2}{cc}\int\int\frac{uu'+vv'+ww'}{r}\,dS\,dS',$$

and this integral extends over all the elements dS and dS′ of conductors S and S′. If, instead of the specific current intensities, one introduces the product from the velocities into the specific densities, and then takes the masses contained in the specific densities as their product and introduces this into the element of volume, then this expression changes to

$$\Sigma\Sigma\frac{\varepsilon\varepsilon'}{cc}\frac{1}{r}\frac{dd'(r^2)}{dt\,dt},$$

if the change in r^2 during time dt which originates in the motion of ϵ is denoted by d, and the change which originates in the motion of ϵ' is denoted by d′.

By disregarding

$$\frac{d\Sigma\Sigma\frac{\varepsilon\varepsilon'}{cc}\frac{1}{r}\frac{d'(r^2)}{dt}}{dt},$$

which disappears through summation for ϵ, this expression can be turned into

$$-\Sigma\Sigma\frac{\varepsilon\varepsilon'}{cc}\frac{d\left(\frac{1}{r}\right)}{dt}\frac{d'(r^2)}{dt}$$

and this, through the addition of

$$\frac{d'\Sigma\Sigma\frac{\varepsilon\varepsilon'}{cc}rr\frac{d\left(\frac{1}{r}\right)}{dt}}{dt},$$

which becomes zero through summation for ϵ', changes into

$$\Sigma\Sigma\varepsilon\varepsilon'\frac{rr}{cc}\frac{dd'\left(\frac{1}{r}\right)}{dt\,dt}.$$

The Derivation of This Law From the New Theory

According to the assumption up to now about electrostatic effects, the potential function U of arbitrarily distributed electrical masses is determined by the condition that

$$\frac{\partial^2 U}{\partial x^2} + \frac{\partial^2 U}{\partial y^2} + \frac{\partial^2 U}{\partial z^2} - 4\pi\varrho = 0,$$

and by the condition that U is continuous and is constant at an infinite distance from the active masses if q denotes its density in points (x, y, z). A particular integral of equation

$$\frac{\partial^2 U}{\partial x^2} + \frac{\partial^2 U}{\partial y^2} + \frac{\partial^2 U}{\partial z^2} = 0,$$

which generally remains continuous outside of points (x′, y′, z′), is f(t)/r and this function forms the potential function which is generated from point (x′, y′, z′) if mass -f(t) is in the same point at time t.

Instead of this, I will now assume that potential function U is determined by the condition

$$\frac{\partial^2 U}{\partial t^2} - \alpha\alpha\left(\frac{\partial^2 U}{\partial x^2} + \frac{\partial^2 U}{\partial y^2} + \frac{\partial^2 U}{\partial z^2}\right) + \alpha\alpha 4\pi\varrho = 0$$

so that the potential function which is generated from point (x′, y′, z′) equals f[t-(r/α)]/r if mass -f(t) is in the same point at time t.

If one denotes the coordinates of mass ϵ at time t by x_t, y_t, z_t and the coordinates of mass ϵ' at time t′ by x_t', y_t', z_t', and if one uses the abbreviation

$$\left((x_t - x'_{t'})^2 + (y_t - y'_{t'})^2 + (z_t - z'_{t'})^2\right)^{-\frac{1}{2}} = \frac{1}{r(t, t')} = F(t, t'),$$

then, according to this assumption, the potential of ϵ on ϵ' at time t

$$= -\varepsilon\varepsilon' F\left(t - \frac{r}{\alpha}, t\right).$$

Therefore, the potential of the forces exerted by all of the ϵ mass of conductor S on the ϵ' mass of conductor S′ from time 0 up to time t will be

$$P = -\int_0^t \Sigma\Sigma\varepsilon\varepsilon' F\left(t - \frac{r}{\alpha}, \tau\right) d\tau,$$

with the sums extended over all the masses of both conductors.

Because the motion for opposite electric masses is an opposing motion in every particle of the conductor, function $F(t,t')$ has the property of being able to change its signs with ϵ through the derivation according to t, and with ϵ' through the derivation according to t'. Then, given the assumed distribution of electricity, if one denotes the derivations according to t by superscripts, and the derivations according to t' by subscripts $\Sigma\Sigma\epsilon\epsilon' F^{(n)}(\tau, \tau)$, which is extended over all the electric masses, does not become infinitely small in comparison to the sum which only extends over one kind of electric mass if n and n' are both uneven.

We will now assume that, during the force's propagation period from one conductor to another, the electric masses cover only a very small path, and we will consider the effect during a time period in contrast to which the propagation period becomes insignificant. In the expression for P, one can first of all replace

$$F\left(\tau - \frac{r}{\alpha}, \tau\right)$$

by

$$F\left(\tau - \frac{r}{\alpha}, \tau\right) - F(\tau, \tau) = -\int_0^{\frac{r}{\alpha}} F'(\tau - \sigma, \tau)\, d\sigma$$

because $\Sigma\Sigma\epsilon\epsilon' F(\tau, \tau)$ can be neglected. In this way, one obtains

$$P = \int_0^t d\tau\, \Sigma\Sigma\varepsilon\varepsilon' \int_0^{\frac{r}{\alpha}} F'(\tau - \sigma, \tau)\, d\sigma,$$

or, if one reverses the order of integration and replaces $\tau + \sigma$ by τ,

$$P = \Sigma\Sigma\varepsilon\varepsilon' \int_0^{\frac{r}{\alpha}} d\sigma \int_{-\sigma}^{t-\sigma} d\tau\, F'(\tau, \tau + \sigma).$$

If one changes the limits of the inner integral in 0 and t, then what is added through this to the upper limits is the expression

$$H(t) = \Sigma\Sigma\varepsilon\varepsilon' \int_0^{\frac{r}{\alpha}} d\sigma \int_{-\sigma}^{0} d\tau\, F'(t + \tau, t + \tau + \sigma)$$

and the value of this expression for $t=0$ is subtracted at the lower limits. So, we have

$$P = \int_0^t d\tau\, \Sigma\Sigma\varepsilon\varepsilon' \int_0^{\frac{r}{\alpha}} d\sigma\, F'(\tau, \tau + \sigma) - H(t) + H(0).$$

One can replace $F'(\tau, \tau+\sigma)$ by $F'(\tau, \tau+\sigma)-F'(\tau, \tau)$ in this expression because

$$\Sigma\Sigma\varepsilon\varepsilon' \frac{r}{\alpha} F'(\tau, \tau)$$

can be neglected. Through this, one obtains an expression as a factor of $\epsilon\epsilon'$ which changes its signs with ϵ as well as with ϵ', so that its terms do not cancel each other out through summation, which also allows infinitely small fractions of individual terms to be neglected. So, when one replaces $F'(\tau, \tau+\sigma)-F'(\tau, \tau)$ by $\sigma[dd'(1/r)/d\tau d\tau]$ and carries out integration for σ up to a fraction that can be neglected, one gets

$$P = \int_0^t \Sigma\Sigma\varepsilon\varepsilon' \frac{rr}{2\alpha\alpha} \frac{dd'\left(\frac{1}{r}\right)}{d\tau\, d\tau} d\tau - H(t) + H(0).$$

It is easy to see that H(t) and H(0) can be neglected, for

$$F'(t+\tau, t+\tau+\sigma) = \frac{d\left(\frac{1}{r}\right)}{dt} + \frac{d^2\left(\frac{1}{r}\right)}{dt^2}\tau + \frac{dd'\left(\frac{1}{r}\right)}{dt\, dt}(\tau+\sigma) + \cdots,$$

and, consequently,

$$H(t) = \Sigma\Sigma\varepsilon\varepsilon' \left(\frac{rr}{2\alpha\alpha} \frac{d\left(\frac{1}{r}\right)}{dt} - \frac{r^3}{6\alpha^3} \frac{d^2\left(\frac{1}{r}\right)}{dt^2} + \frac{r^3}{6\alpha^3} \frac{dd'\left(\frac{1}{r}\right)}{dt\, dt} + \cdots\right).$$

But, this is only the first term of the factor $\epsilon\epsilon'$ with the factor in the first component of P having the same order, and, because of summation for ϵ', this supplies only a fraction, which can be neglected, of the same term.

The value for P, which results from our theory, agrees with the empirical

$$P = \int_0^t \Sigma\Sigma\varepsilon\varepsilon' \frac{rr}{cc} \frac{dd'\left(\frac{1}{r}\right)}{d\tau\, d\tau} d\tau$$

if we assume that $\alpha\alpha = \frac{1}{2}cc$.

According to *Weber* and *Kohlrausch's* definition, $c = 439450 \cdot 10^6$mm/sec, which results in α being 41949 geographical miles per second, while *Busch* found the speed of light to be 41994 miles from *Bradley's* aberration observations, and *Fizeau* found it to be 41882 miles by direct measurement.

This paper was presented by *Riemann* to the Royal Society of Sciences at Göttingen on February 10, 1858, according to a note added to the title of the manuscript by the Secretary of the Society at that time, but it was later retracted. After the paper had been published after *Riemann's* death, it was subjected to criticism by *Clausius* (*Poggendorff's* Annalen Bd. CXXXV p. 606), whose essential objection consists of the following.

According to the provisions, sum

$$P = -\int_0^t \Sigma\Sigma\varepsilon\varepsilon' F\left(\tau - \frac{r}{\alpha}, \tau\right) d\tau$$

has an infinitesimally small value. Therefore, by virtue of a noninfinitesimally small value that will be found later for it, the operation must contain a mistake, for *Clausius* found an unjustified reversal of the sequence of integration in the exposition.

The objection appears to me to be well-founded, and I am of *Clausius's* opinion that *Riemann* himself discovered the same and retracted the work before publication because of this.

Although the most essential content of *Riemann's* deduction collapses as a result of this, I have decided to include this treatise in the present collection because I did not want to presume to decide whether it nevertheless did not still contain some kernels of further fruitful thought on this highly interesting question.

Weber.

Index

S

T

U

V

W

Y

Z